MESSAGES IN STONE

Colorado's Colorful Geology

COLORADO GEOLOGICAL SURVEY

Edited by Vincent Matthews, Ph.D., Katie KellerLynn, & Betty Fox

Contributing Authors

James A. Cappa, Gary Curtiss, Tom Hemborg, John W. Keller, Katie KellerLynn, Susan Landon, Vincent Matthews, Matt Morgan, David C. Noe, Pat Rogers, Matthew Sares, Beth Widmann, Laura Wray, & Knox Williams

Colorado Geological Survey
1313 Sherman Street, Room 715
Denver, Colorado 80203

General: 303-866-2611
Publications: 303-866-4762
www.geosurvey.state.co.us
cgspubs@state.co.us

ISBN 1-884216-59-5

Printed in Canada

Cover image by Francisco Gutierrez: Fracturing in Jurassic sandstones on the eastern flank of the Paradox Valley, Colorado

Acknowledgements

We are grateful for the interest, generosity, and contributions of a great number of people. We particularly would like to thank Ralph Lee Hopkins, John A. Karachewski, Theodore Walker, Sandra J. Lindquist, Wallace R. Hansen, P. J. Hasselbach, Mark Wark, Dave Catts, Ray Troll, and Kirk R. Johnson for helping to make this a well-illustrated book. Susan Hill Newton served as managing editor and Rick Ciminelli of the state's IDF Design Center devoted his talents to the design of the book.

We are grateful to our reviewers who significantly improved the manuscript: Eric Nelson and Maeve Boland, Colorado School of Mines; Jim McClurg, University of Wyoming; Jack Stanesco, Red Rocks Community College; Pete Modreski, United States Geological Survey; Chris Hayes, Attorney at Law; Vicki Cowart, former State Geologist of Colorado; and Emmett Evanoff, University of Colorado. Richard Madole, United States Geological Survey Emeritus, reviewed the glacial and wind-blown sections and Kirk R. Johnson, the Denver Museum of Nature & Science, reviewed the fossils section.

PHOTOGRAPHS AND IMAGES CONTRIBUTED BY:

Individuals

Peter Birkeland	Richard Madole
Bruce Bryant	Vincent Matthews
Dave Bunk	Bill Middlebrook
James A. Cappa	Karen Morgan
Chris Carroll	Matt Morgan
Dave Catts	Eric Nelson
William A. Cobban	David C. Noe
Howard Coopersmith	Brian Penn
Emmett Evanoff	Richard M. Pratt
Mike Frasier	Jerry Roberts
David A. Gonzales	W. P. Rogers
Francisco Gutierrez	Eric Route
Wallace R. Hansen	Matthew Sares
David Harris	Larry Scott
P. J. Hasselbach	Jeff Scovil
Ralph Lee Hopkins	Kenneth Lee Shropshire
Kirk R. Johnson	James Soule
John A. Karachewski	Jack Stanesco
Allen Karsh	Ray Troll
John W. Keller	Joe Tucciarone
G. Kent Keller	Theodore Walker
Katie KellerLynn	Mark Wark
Brendan LaMarre	Jonathan White
Trina LaMarre	Beth Widmann
Tim Lane	Jason Wilson
Sandra J. Lindquist	Laura Wray
Peter Lipman	

Institutions

American Soda
Colorado Department of Transportation
Colorado Historical Society
Denver Museum of Nature & Science
Florissant Fossil Beds National Monument
National Aeronautics and Space Administration
Nevada Bureau of Mines and Geology
Rocky Mountain National Park
Stagner, Inc.
United States Army
United States Geological Survey
University of Colorado, Department of Geological Sciences

Preface

THE PLAINS, MOUNTAINS, VALLEYS, AND PLATEAUS—indeed every aspect of the natural landscape—are the direct result of geologic processes. Cataclysmic forces have pushed, pulled, hammered, and baked the state into the ever-changing scenery we call Colorado, the Centennial State.

Written for amateur and professional geologists alike, *Messages in Stone* reflects a firmly held conviction steeped in knowledge. Simply by understanding the forces that create, and continue to shape, the land we explore and live in, your pleasure will be enhanced and your ability to live successfully with its complex and challenging geology will increase.

—Ronald W. Cattany
Interim State Geologist and
Director of the Division of Mines and Geology

Contents

Introduction

As we, the scientists of the Colorado Geological Survey (CGS), travel around Colorado carrying out our mission, we are fortunate to be able to observe the diversity of breathtaking beauty displayed in every part of the state. The geologic processes and history that created, and that are still creating, this magnificent landscape are world class. We are fortunate to have excellent examples of most of the classic features of geology, and many geologic processes can be observed in action today.

The beauty of Colorado's scenery is easily awe-inspiring; but being equipped with knowledge to figure out the possible properties and nature behind it all only increases your enjoyment. The CGS mission is to serve and inform the people of Colorado by providing sound geologic information and evaluation, and to educate the public about the important role of earth sciences in everyday life in Colorado. *Messages in Stone* is intended to do just that. We wish to share with you some insights into the exciting geologic features and history that create our scenery, provide us with vast mineral resources, and create potentially hazardous conditions that must be respected.

Setting sun on Cretaceous sandstones of Mesa Verde in the foreground, with the igneous and sedimentary rocks of the La Plata Mountains in the background

Volcanic breccias of The Castles in the West Elk Mountains. In 1876, a USGS geologist wrote, "The hills west of Ohio Creek are composed mainly of breccia ... eroded in the most fantastic fashion. The breccia is stratified, and there are huge castle-like forms, abrupt walls, spires, and towers."

Facing page: Aerial view of star dunes in Great Sand Dunes National Park

The spectacular, diverse scenery in many parts of the state is directly related to its colorful, varied geology. The jagged, sedimentary layers of the Maroon Bells create a totally different feel and look than the rounded, granitic rocks of Rocky Mountain National Park. The badlands of shale around Delta contrast sharply with the lofty volcanic plateau of Grand Mesa. The craggy volcanic rocks of southwestern Colorado's "Switzerland of America" couldn't be more different than the windswept sand dunes of the eastern plains. The Arkansas River carves 1,000 feet into the ancient metamorphic rocks of Royal Gorge, but in a short distance meanders lazily past the Teepee Buttes of the Great Plains. Each geologic area has its own beauty and each has its own story to tell.

Right: Tertiary igneous dikes radiating from West Spanish Peak

Below: Gray Tertiary sills intrude red sedimentary rocks of the Pennsylvanian Maroon Formation in Jacque Peak southwest of Copper Mountain

Permian sandstones of Garden of the Gods

For instance, the striking salmon-colored spires of Garden of the Gods in Colorado Springs, give us a taste of how much geologic history can be read in one place. The rocks let us know that a couple of hundred million years ago this area was a Middle-Eastern-type desert with sweeping dunes along coastal mountains. The fact that the layers are standing on end tells us much about the way mountains were built 60 million years ago. The surfaces that cut across the layers bare the enduring scars of climate conditions from hundreds of thousands of years ago during the ice ages.

But Garden of the Gods is just one of the thousands of examples. So get out, enjoy and admire, and educate yourself on Colorado's surrounding natural history.

Precambrian igneous and metamorphic rocks in Unaweep Canyon

Devil's Backbone

The colorful nature of Colorado extends to its history and names. For example, many treacherous topographic features in Colorado are named after the devil. It is interesting that all of these namesakes are found in only two types of rock, either igneous rock or Mesozoic sandstones. Tortured topographic forms found in igneous rock include: Devil's Head, which can be seen on the skyline to the southwest of the Denver metropolitan area; Devil's Kitchen on the Rampart Range; Devil's Gulch southwest of Westcliffe; Devil's Knob south of Saguache; Devil's Lake and Devil's Creek northeast of Lake City; Devil's Lookout in Black Canyon of the Gunnison National Monument; Devil's Nose southwest of Idaho Springs; Devil's Slide southwest of Colorado Springs; Devil's Causeway in the Flattop Wilderness; and Devil's Stairsteps south of La Veta.

The next group of demonic features are scattered across the state but form in Mesozoic sandstones that tend to break into rectangular jumbled masses, including: Devil Mountain east of Bayfield; Devil Point south of the Dolores River; Devil's Elbow in Ute Canyon near the Oklahoma border; Devil's Kitchen in Colorado National Monument; Devil's Grave Mesa northwest of Yampa; Devil's Backbone west of Loveland; Devil's Hole northwest of Meeker; Devil's Rocking Chair east of Ninaview; Devil's Rockpile east of Marble; and Devil's Stairway along Anthracite Creek.

Upper left: Devil's Backbone west of Loveland. Many initially suspected that this was an igneous dike until inspection revealed that it is a vertical bed of sandstone, which is part of the steep limb of an anticline.

Above: Devil's Stairsteps were formed by an igneous dike radiating from West Spanish Peak.

Below: Devil's Head, visible from Denver, is a ragged, rounded mass of resistant granite that is raised above the surface of the Rampart Range by faulting.

MESSAGES IN STONE

Colorado's Gee-whiz-ology

Colorado has magnificent geology that is beautifully displayed for all to see. The state holds many of the biggest, the best, the first, and the most diverse. For instance, did you know…

- The Rocky Mountains have fifty-eight peaks more than 14,000 feet high, all of them in Colorado.
- Colorado has the highest average elevation of any state (6,800 feet), with more than two vertical miles between its lowest and highest points.
- There are more than 774 different types of minerals found in Colorado.
- The largest faceted diamond produced in the United States came from Colorado.
- Colorado has more than 30,000 square miles of wind-blown deposits and virtually every kind of sand dune, including fossil dunes.
- The very first dinosaur discovered in western North America and the first *T. rex* in the world were both found in Colorado.
- Colorado has more than a dozen major dinosaur fossil sites.
- The bones of three of the four largest dinosaurs found in the world were found in Mesa County.
- The world's first Brontosaurus (*Apatosaurus*) tracks were recognized in Colorado, and the continent's longest dinosaur trackway site is in the southeast corner of the state.
- One of the most diverse and oldest fossil tropical rainforest leaf assemblages in the world was found near Castle Rock.
- The world's only fossil tsetse flies, and fossils of the world's oldest roses, cheetahs, and ferrets were found in Colorado.
- Two of the world's largest molybdenum deposits and the world's largest titanium deposit are located in Colorado.
- Colorado has the largest oil shale resources in the United States.

Mount Elbert is the highest point in Colorado (and the Rocky Mountains) at 14,433 feet above sea level. It looms one mile above the Twin Lakes at its base. The lakes are dammed by glacial moraines.

- The second oil field in the United States was discovered in Colorado.
- More than half of Colorado's counties produce oil and gas, and more than 55,000 wells have been drilled for oil and gas in Colorado.
- A suburban area southwest of Denver has the highest annual monetary loss in the nation from swelling soil.
- Colorado has the greatest number of avalanches in the lower forty-eight states and the most avalanche-related deaths in the United States.
- Every kind of anticline found in Earth's crust can be observed in Colorado.
- Colorado has nineteen calderas, including one of the largest in the world.
- Colorado has one of the world's largest ash flow sheets whose volume is greater than 1,200 cubic miles.

- Colorado has real glaciers in the Front Range and some of the most spectacular rock glaciers in the world.
- Colorado marble was used in the Lincoln Memorial and the Tomb of the Unknowns in Washington, D.C.
- Madame Curie won a Nobel Prize using Colorado uranium.
- Colorado has a Cretaceous/Tertiary (K/T) site with the highest iridium content ever measured in continental rocks.
- Colorado is the birthplace of the Rio Grande, Colorado, North and South Platte, Arkansas, San Juan, Dolores, Gunnison, and Yampa Rivers, and its rivers flow to eighteen other states.
- Colorado experienced a M 6.6 earthquake in 1882, has more than ninety Quaternary faults, and is the world's outstanding laboratory for human-caused earthquakes.
- The oldest rock in Colorado is 2.7 billion years old, the youngest volcanic rock is only 4,150 years old, and some rocks in Colorado came from 100 miles deep in the mantle.
- Some of the metamorphic rocks in Colorado's mountains were formed at pressures of sixty-seven tons per square inch, at 1,000 degrees Fahrenheit.

DISTRIBUTION OF PEAKS MORE THAN 14,000 FEET IN ELEVATION

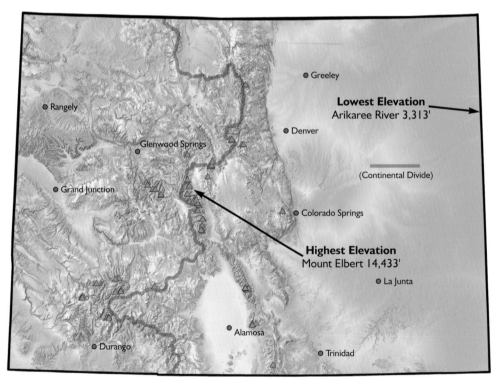

Greeley

Rangely

Glenwood Springs

Denver

Lowest Elevation
Arikaree River 3,313'

(Continental Divide)

Grand Junction

Colorado Springs

Highest Elevation
Mount Elbert 14,433'

La Junta

Alamosa

Durango

Trinidad

Note that the eastern third of Colorado is relatively flat and featureless and how abruptly the mountains rise from the plains.

How many fourteeners? The question of how many 14,000+ foot peaks are in Colorado depends on whom you ask. The "fourteeners" website claims fifty-three, Colorado Mountain Club claims fifty-four, and the book *Colorado's Fourteeners* claims fifty-five.

USGS has the official responsibility for measuring, mapping, and naming topographic features in the United States. They list, by name, fifty-eight summits in Colorado that have elevations more than 14,000 feet above sea level.

COLORADO PHYSIOGRAPHIC PROVINCES

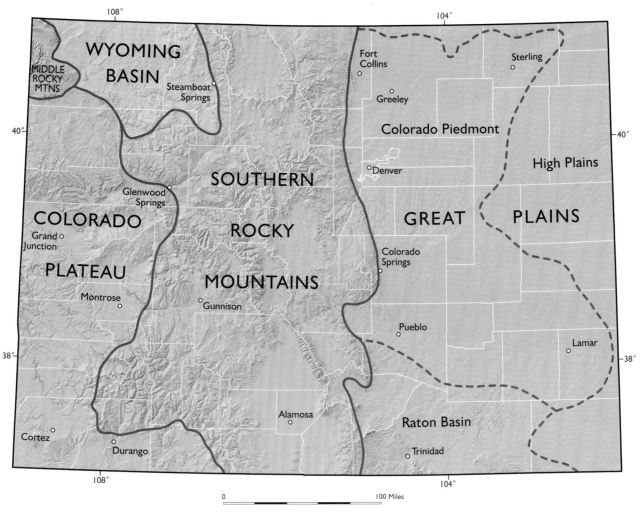

Colorado's setting includes five physiographic provinces—a region in which the geologic structure, climate, and relief and/or landforms are significantly different from that of adjacent regions. These provinces are the Great Plains, Southern Rocky Mountains, Middle Rocky Mountains, Colorado Plateau, and Wyoming Basin. The Great Plains province is subdivided into the High Plains, Colorado Piedmont, and Raton Basin (outlined on the map by dashed lines). From Fenneman, 1946.

"On all the broad extent of these United States, certainly no region can be found which presents more facts of interest, more opportunities for investigation, and greater possibilities of new discoveries, than the state of Colorado."

—S. F. Emmons, 1883, United States Geological Survey
From the presidential address to the inaugural meeting of the Colorado Scientific Society

Major geographic place names in Colorado

1. Arkansas Graben
2. Black Canyon of the Gunnison
3. Blue River Graben
4. Book Cliffs
5. Elk Mountains
6. Elkhead Mountains
7. Front Range
8. Gore Range
9. Grand Hogback
10. Grand Mesa
11. Grand Valley
12. Great Sand Dunes
13. LaPlata Mountains
14. Medicine Bow Mountains
15. Middle Park
16. Mosquito Range
17. North Park
18. Paradox Valley
19. Palmer Divide
20. Park Range
21. Pawnee Buttes
22. Pike's Peak
23. Purgatoire Canyon
24. Rabbit Ears Range
25. Rampart Range
26. Roan Plateau
27. Sand Wash Basin
28. Sangre de Cristo Range
29. San Juan Mountains
30. San Luis Valley
31. Sawatch Range
32. Sleeping Ute Mountain
33. South Park
34. Spanish Peaks
35. Tarryall Mountains
36. Ten Mile Range
37. Two Buttes
38. White River Uplift
39. Uinta Mountains
40. Uncompahgre Plateau
41. West Elk Mountains
42. Wet Mountains
43. Williams Fork Mountains
44. Wray Dune Field

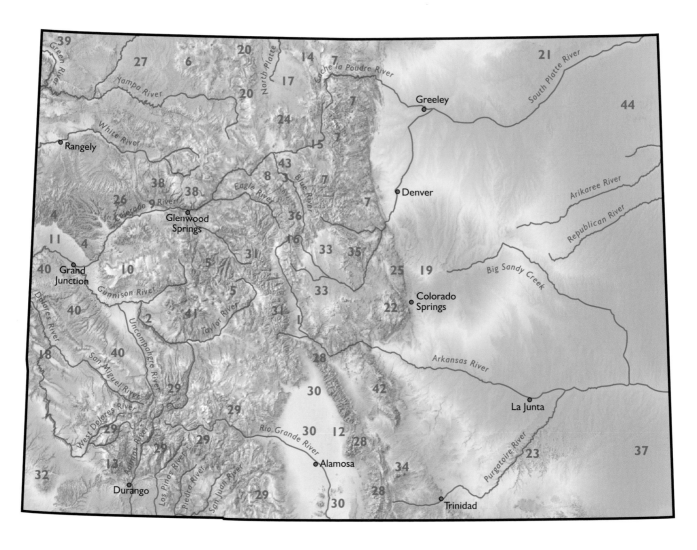

CHAPTER 1

Rocks and Contortions: Colorado's Rocks, Structures, and Maps

Pick up a rock. Tracing its striations and indentations with your finger, you feel its ancient wounds resulting from having been buried deep below Earth's surface, compressed by tons of material, heated until it became partly liquid, and folded as it cooled. Most amazing is that miles of overlying rock may have been removed to bring it to the surface. You will find other rocks that are a hodge-podge of many types of rocks that have been moved, reformed, and finally pressed together to become the stone you hold in your hand.

Rocks are priceless recorders. The stories locked in the rocks recount mountain ranges rising time and again only to be buried in their own debris, shallow seas sweeping across the land, and deserts undulating with massive dunes. The strata tell of erupting volcanoes that seared the land with lava and filled the air with white-hot ash, and of a land dominated by coastal deltas and swamps. Etched in the rocks is an echo of glacial ice more than half a mile thick, scraping off mountain peaks and scouring valleys in its slow, inexorable progress.

Geologists labor to unravel the clues and reassemble the history of Earth's distant, tumultuous past by studying rocks and their relationship to one another. To piece together the story, geologists must assume many roles—historian, detective, explorer, chemist, physicist, and mathematician. These renaissance scientists use many techniques and classification systems to categorize rocks.

The purpose of naming types of rocks is to indicate specific chemical compositions as well as the shape, character, and proportions of the minerals they contain. Contrary to popular belief, professional geologists don't just stroll out, pick up a rock at random, peer at it, and rattle off its forty-six-syllable classification. To be truly accurate in their analysis, scientists carry samples back to laboratories where rocks are cut (thinner than a piece of paper), stained, crushed, and subjected to all sorts of tests including chemical and X ray.

Unfortunately—although it is common knowledge, at least among geologists, that they are exceptionally strong, nimble, intelligent, and attractive examples of humankind—they can't always hoist a boulder onto their shoulder and stagger off to the nearest laboratory. So, they devised a simpler first-level classification system that can be used in the field. It uses visual clues such as texture, color, and grain size.

But for the novice geologist, being able to differentiate among three broad groups—igneous, sedimentary, and metamorphic—will easily help in understanding and appreciating Colorado's scenery. Igneous rocks crystallize from molten material. Simply put, when rocks break down from weathering on the surface, the fragments are carried away and deposited as sedimentary rocks. Applying heat and pressure to either igneous or sedimentary rocks recrystallizes them into metamorphic rocks.

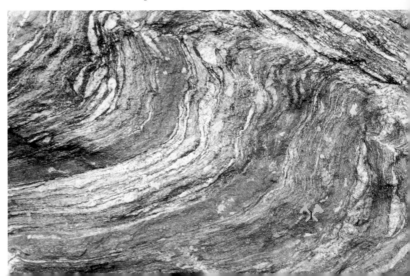

1

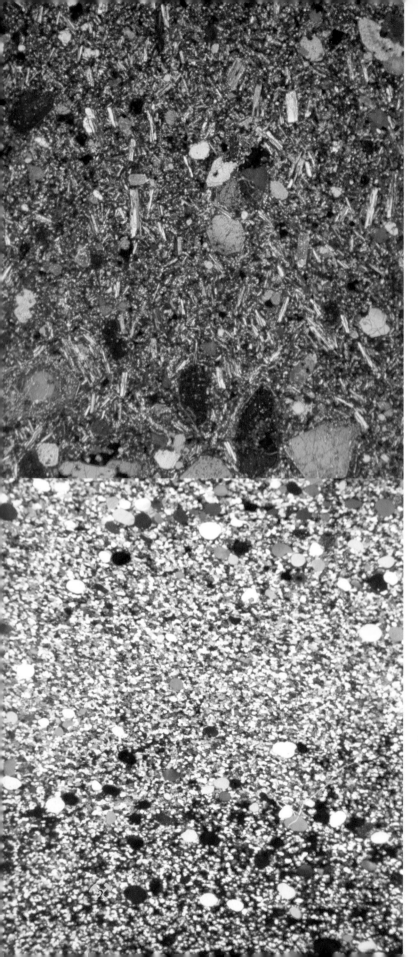

Land of Fire: Colorado's Igneous Rock

Earth's oldest rocks, and those from which the other two types of rock are ultimately derived, are igneous, from the Latin word for fire. Colorado has igneous rocks of many ages, shapes, sizes, and compositions including some rare and exotic ones. Because they originated as molten material pushing between, or cutting through, other rocks, the displays they create are often strikingly dramatic.

Molten rock is extraordinarily hot, sometimes exceeding 2,000 degrees Fahrenheit. If this molten and partially crystallized material (magma) crystallizes underground before it reaches the surface, the rock is "intrusive" or "plutonic." If magma travels up through Earth's crust and reaches the surface, the resulting rocks are "extrusive" or "volcanic."

Geologists have learned that the more iron and calcium in a rock, the darker the color. Light-colored igneous rocks contain more silicon, potassium, and sodium. Combining clues about the chemical composition found in the color of igneous rocks with an observation of the coarseness of grains makes further distinctions possible. For example, coarser-grained crystalline plutonic rocks are divided into "granitic" or "gabbroic," depending on whether they are light colored or dark colored, respectively. Finer grained volcanic rocks are divided into rhyolite, andesite, or basalt because they are light colored, medium colored, or dark colored.

The image above is igneous rock, and the one below is sedimentary rock. These two rocks were sliced to less than $1/1,000$ of an inch thick. The thin sections are then viewed through a special microscope using polarized light to understand the texture and mineralogy of the rocks. Note that the minerals in the igneous rock have straight boundaries with angular bends, whereas the minerals in the sedimentary rock have discrete grains with rounded edges. This shows a geologist that the rock above crystallized from molten material, and the rock below is sandstone formed from grains rounded by rivers or wind.

PLUTONIC OR INTRUSIVE ROCK

Magma that moves upward and cuts across pre-existing layers of rock forms what are generally known as plutons. The largest plutons are batholiths, which are composed of granitic rocks that, by definition, exceed forty square miles in area. The granitic rocks of Pike's Peak are part of a 1,300-square-mile batholith, and the batholith surrounding Estes Park covers more than 600 square miles. These batholiths solidified deep in the crust, but are now at the surface because miles of overlying rock were removed by erosion.

Smaller plutons take a variety of shapes, each with its own name. Stocks are irregular bodies much smaller than batholiths; dikes are tabular, sheet-like bodies that cut across the bedding or foliation of the host rock; and plugs are necks of solidified igneous rocks. Sills differ from most other intrusions because they intrude between, and parallel to, the layers of the host sedimentary rock. Laccoliths appear to have started as sills, but, as more magma intruded, they pushed up the overlying layers rather than spreading out laterally and formed mushroom-shaped bodies with flat floors and domed roofs. Colorado has many laccoliths and is the first place in the world where they were recognized and named. A large field of laccoliths extends from Mount Sopris south of Glenwood Springs for forty miles.

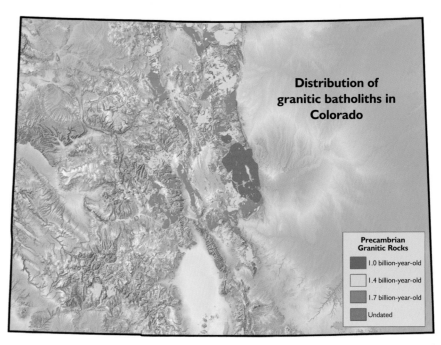

Distribution of granitic batholiths in Colorado

Precambrian Granitic Rocks

- 1.0 billion-year-old
- 1.4 billion-year-old
- 1.7 billion-year-old
- Undated

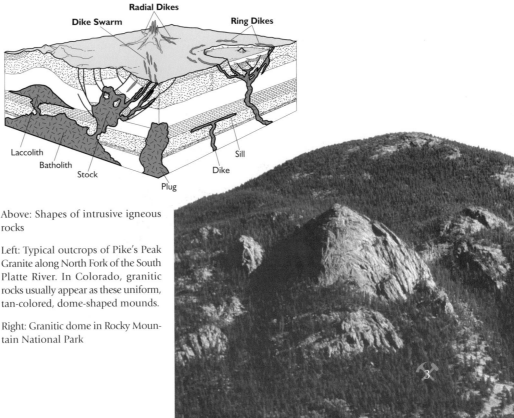

Radial Dikes

Dike Swarm

Ring Dikes

Laccolith

Batholith

Stock

Plug

Sill

Dike

Above: Shapes of intrusive igneous rocks

Left: Typical outcrops of Pike's Peak Granite along North Fork of the South Platte River. In Colorado, granitic rocks usually appear as these uniform, tan-colored, dome-shaped mounds.

Right: Granitic dome in Rocky Mountain National Park

Upper left: Geologic map of radiating dikes at Dike Mountain. Red blob in center is a plug and the red lines are dikes radiating outward.

Above: These dikes can be observed radiating out from the central intrusion at the top of Dike Mountain along Highway 164, twenty-five miles west of Walsenburg.

Left: Dikes radiating from West Spanish Peak

Below: Vertical Tertiary dike intruding Tertiary sedimentary strata southwest of Trinidad

Upper left: Tertiary Pando Porphyry intrudes lower Paleozoic sedimentary rocks on the north slope of Mount Sheridan

Above: Near the Green Mountain Reservoir Dam, dikes and sills (light colored) are beautifully exposed intruding Cretaceous shales (dark colored). The cross-cutting dike appears to feed the sill and the overlying thick laccolith. Only the floor of the laccolith is visible in this photograph.

Left: Brown Tertiary sills intruding black Tertiary shales west of Trinidad. The brown sills at first glance resemble the Tertiary sandstones in the area.

Below: At least nine sills (lighter colored) exposed near the summit of 13,232-foot-high Hesperus Peak in the La Plata Mountains. The sills intrude Cretaceous shales (brown) and increase the overall volume in the Hesperus Peak by 40 percent or 1,000 feet.

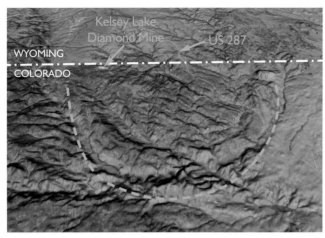

Digital elevation model of Virginia Dale ring dike. The circular structure is ten miles across and is located in northern Colorado along Highway 287.

Tertiary plug near Gardner

A gabbro plug along I-25 north of Walsenburg at Huerfano Butte

Many have incorrectly concluded that the Spanish Peaks are old volcanoes. In fact, almost the entire mountain is comprised of Tertiary sedimentary rocks with only a small intrusion near the crest. The abundant dikes, radiating in all directions from the mountain, have been studied by geologists all over the world. In 1869, Dr. F. V. Hayden of the USGS first examined the Spanish Peaks and described them as a "gigantic dike."

6

Chair Mountain laccolith intrudes Cretaceous shales

South of Carbondale, 12,660-foot-high Mount Sopris bows up Pennsylvanian shales and evaporites

Needle Rock plug east of Crawford is more than 800 feet high. Tater Heap laccolith has baked shales bowed up along its margins.

Location of laccoliths in west-central Colorado. Other laccoliths are scattered around the state.

Gothic Mountain laccolith and sill intrudes Cretaceous shales

Needle Rock Plug

Tater Heap Laccolith

Crested Butte

Gunnison

Laccolith

Sill

Cretaceous Shales

Rocks and Contortions

7

Volcanic or Extrusive Rock

Colorado has volcanic rocks of many ages and origins. Magma doesn't always solidify underground. Often some of it reaches the surface where it forms a variety of volcanic landforms and deposits.

Basaltic lava flowed out on the ground many times in Colorado's past and often covered large areas. When the lava cooled and solidified, it formed dense, hard, resistant basalt. Columnar jointing commonly formed as the rock cooled and contracted. Today, these resistant volcanic flows cap mesas such as Grand Mesa, White River Plateau, Raton Mesa, and those near Basalt.

Above: Fisher Peak, looming over Trinidad, is part of the basalt-capped Raton Mesa and displays at least four distinct flows with columnar jointing. Inset: A basalt dike intruding sedimentary rocks along I-25 south of Trinidad. This is one of the dikes that fed the lava flows above on Raton Mesa.

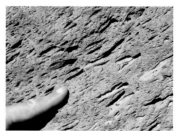

Gas bubbles trapped in solidifying lava elongate by continued movement and become stretched vesicles in the rock.

The basalts can usually be identified by their dark color and columnar jointing. Unfortunately, there are not any classic cinder-cone edifices left in Colorado because they have been mined by humans or eroded by nature. Mesita cinder cone in the San Luis Valley has been completely leveled and distributed around the country to be used in barbecue grills and decorative rock gardens. However, there are remnants of cinder cones in several places in central Colorado. Ancient basaltic shield volcanoes are found west of Antonito and north of Basalt.

Left: Close-up of columnar jointing near North Pass on CO-114 that commonly forms as basalt, or andesite magma cools, and contracts at the surface or very shallow underground.

These columns formed in basalt-like flows of shoshonite on South Table Mountain as the lava cooled and contracted.

Close-up of West Elk Breccia. Excellent exposures of this huge deposit are on the north side of Blue Mesa Reservoir.

But to appreciate Colorado's lively volcanic past you need to understand ash, breccia, tuff, and calderas. When a volcano erupts violently, fragments of rock, pumice, and lava are blown into the air. The fragments range from fine dust to large blocks. Fragments the size of sand are ash. When ash hardens into rock, it becomes tuff. Rocks composed of large angular fragments of volcanic material are breccia.

Large, angular blocks in a volcanic breccia along the south flank of Rabbit Ears Pass in northern Colorado

When a volcano blows ash up into the air—sometimes into the stratosphere—it produces ash falls. The ash cools and eventually falls out as a blanket on the landscape, forming cold deposits. An ash flow is also the product of a volcanic eruption, but, rather than being carried away and cooled by the winds, the hot particles and gas fall back to the ground and spread out laterally as a hot, glowing cloud of ash and gas. The gas trapped between the hot particles acts as a lubricant and enables the material to flow fifty to seventy-five miles at speeds of more than seventy-five miles per hour. When the ash flow comes to rest, the particles at the bottom of the deposit are often still quite hot and ductile. If conditions are just right, the overlying weight of the deposit compresses the hot, pliable particles, collapses the holes in the pumice, and welds the mixture into a dense, glassy rock called a welded tuff.

Eruption of large ash flows evacuates the magma chamber, causing the surface to collapse and a caldera to form. The ash cloud is composed of particles blown high into the atmosphere. These particles will cool and drift off to eventually be deposited as ash falls. Some large eruptions deposit ash falls worldwide. The hot particles of the ash flow fall back to the ground while still hot, and then rush out over the countryside at speeds of up to seventy-five miles per hour.

Colorado has ash-flow tuffs that cover thousands of square miles in the southwestern part of the state. The Fish Canyon Tuff, surrounding Creede, is one of the largest in the world. It contains nearly 1,200 cubic miles of material and has a welded zone more than a half-mile thick.

Ash flow eruptions and caldera formation are closely related. When large ash flows erupt, the roof collapses into the large, subsurface void created by the removal of massive amounts of material, and a roughly circular depression (caldera) is formed. The eruption that caused the Fish Canyon Tuff also created the enormous La Garita caldera measuring twenty-two miles wide and forty-seven miles long. There are at least nineteen calderas in Colorado making the state one of the world's best outdoor laboratories in which to study their formation.

The white layer is an ash-fall tuff in Cretaceous rocks at Dinosaur Ridge. The ash probably came from volcanic eruptions in Utah.

MESSAGES IN STONE

Artist's conception of the eruption of the Wall Mountain ash flow from the vicinity of the present day Sawatch Range. The hot, incandescent cloud of ash and gas raced across central Colorado to the vicinity of Castle Rock in about an hour, consuming everything in its path. As this huge volume of material erupted, the surface foundered into the magma chamber.

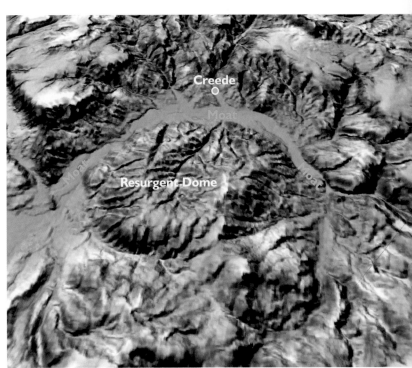

Creede has one of the most pristine examples of a resurgent dome in a caldera. The caldera has been thoroughly studied by detailed mapping and extensive drilling. The moat circles the outside of the resurgent dome and lies within the caldera walls. Calderas commonly have lakes filling their collapsed depressions, such as Crater Lake National Monument in Oregon. The moat at Creede is underlain by sediments from ancient Lake Creede.

Outcrops of the 1,200-cubic-mile deposit of Fish Canyon Tuff are well displayed along Highway 160 southwest of South Fork.

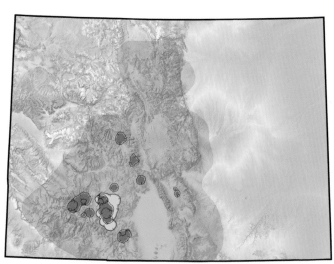

Interpreted original extent (light red) of middle Tertiary volcanic deposits in Colorado. Red are middle Tertiary calderas; tan is the La Garita caldera

Beautiful exposures of the roots of the Cimarron volcano in the San Juan Mountains about eight miles northeast of Ouray. The horizontal layers are a variety of volcanic deposits. The dikes that fed the now-eroded volcano cut upward across the horizontal layers. Inset: Geologic map showing the central plug and radiating dikes of the Cimarron volcano.

These tuffs in the Wheeler Geologic Area show varying degrees of welding. The light slopes are not very welded and thus are highly erodible. The darker rocks are more densely welded and thus resist erosion, protecting the softer tuffs below. The tuffs erupted during the formation of the San Luis caldera.

"Castle" of Wall Mountain tuff, a 37-million-year-old ash flow. This flow traveled ninety miles from its source in the Sawatch Range to the vicinity of Castle Rock on the eastern plains. The outcrop shown is in Castle Rock Gulch, east of Buena Vista.

COLORADO'S METEORITES

At the extreme of Colorado's igneous rocks are the extra-terrestrial meteorites. Look up on a clear night and you may be lucky enough to see a "shooting star" blaze across the sky. What you are probably seeing are pieces of rock that enter Earth's atmosphere. Meteorites originate from celestial bodies, such as asteroids and planetary surfaces, and some even contain material remaining from the formation of our solar system.

It wasn't until late in the 1920s that the fledgling field of meteoritics began to be recognized as more than just a topic of interesting conversation and after-dinner dialogue. As the young science began to spread its wings, much of its focus, particularly during the first sixteen years, was on Colorado and the Museum of Natural History (now the Denver Museum of Nature & Science) where the most intensive study of meteorites was conducted.

One innovative aspect of the project was to train farmers to watch the sky and to notify museum staff if they found any rocks on their land that they suspected could be meteorites. The success of this effort is evident in the number of significant finds reported. In 1909, there were only four iron meteorites reported in Colorado. By 1930, there were twelve known meteorites in the state. By 1936, that number had grown to a total of twenty, and by 1950 there were forty-six recognized meteorites.

Some meteorite landings have been well observed. A meteorite shower startled the people of Johnstown, Colorado, on the afternoon of July 6, 1924. The following passage is an account of the meteorite shower, as described by an eyewitness:

Artist's conception of a fireball over the Maroon Bells

"At 4:20 in the afternoon on a cloudless day, four terrific explosions were heard. Then came a series of minor bursts, like the crackling of a machine gun. Other accounts describe the noises as resembling 'shrill screeches,' 'whistling,' and 'the exhaust of an airplane.' Grayish blue 'smoke' puffs are described as accompanying the explosions, and then missiles struck with a 'thud,' a 'thug,' or a 'thump.' At least four of the largest fragments were seen to fall, and although they were separated by considerable distances, the course of each was marked by a trail of 'light gray smoke.' The first piece unearthed fell in the highway about thirty feet from the doors of the little church at Elwell. At the time a funeral service was being held in the yard behind the church, and the fall was witnessed and heard by not fewer than two hundred persons."

One of the Johnstown meteorites. This photo is twice the size of the meteorite, which is cut in half exposing the unusual character of the diogenite. On the dark fusion crust is the original number assigned by H. H. Nininger.

Hole in roof caused by a meteorite in 1973 in Cañon City

The Johnstown meteorite was of great scientific importance. Not only was it recovered immediately after it fell, it was also a rare type of diogenite meteorite. These meteorites appear to have cooled slowly as a volcanic dike or sill on their parent body. Scientists believe the diogenites are from the asteroid Vesta.

Colorado has seventy-nine known meteorites—a number exceeded only by Texas, Kansas, and New Mexico. Chemistry, physics, and geology have helped unlock the secrets of meteorites and the clues they contain regarding the birth of the Sun, our planet, and the universe as a whole. Had it not been for H. H. Nininger and the American Meteorite Laboratory in Denver, collections of meteorites in museums around the world would be a fraction of what they are today.

This slice of a meteorite, with a beautiful Widmanstätten structure, was found along Bear Creek in Jefferson County in 1866.

Below: Meteoroid-eye-view looking northwest showing the curvature of Earth. The lower part of the photo shows the northwest quarter of Colorado. Can you find the Great Salt Lake and Yellowstone National Park in the distance?

MESSAGES IN STONE

The Book Cliffs north of Grand Junction are aptly named because much of the story of Colorado's ancient past is recorded in the sedimentary rocks, or pages, of Earth history.

Bits and Pieces: Colorado's Sedimentary Rocks

Blown across the land by wind or carried along by water and ice as the land continued to remake itself, loose sediments eventually compressed and cemented into rock and left messages in stone for us to decipher. Sediments include the mud at the bottom of streams, the sand dunes at the foot of the mountains, the chemical precipitates of salt in shallow seas, the beaches at the edge of inland seas, and the graveyards of tiny fossils at the bottom of tropical oceans. In these sedimentary layers, the imprints of changing life forms in an ancient world are faithfully recorded.

Sedimentary rocks are laid down in layers that vary in thickness from less than an inch to hundreds of feet. The boundaries between successive layers (contacts) are marked by subtle changes in grain size or color, or an abrupt change of rock type. Within the sedimentary layers, other structures give clues to the rocks' origin. Sedimentary structures such as ripple marks, cross bedding, graded bedding, mud cracks, raindrop impressions, and animal tracks pattern the surfaces of many sedimentary layers.

Reading the layers of sediment, geologists are not only able to determine what geologic events took place, but also their order and approximate timing. In Colorado, sedimentary rocks bear witness to a great number of changes in an extraordinary number of environments. Sedimentary materials in Colorado were deposited in shallow lakes, streams, beaches, alluvial fans, dune fields, estuaries, deep and shallow seas, offshore bars, restricted evaporite basins, turbidite fans, sabkhas, algal mounds, washover fans, wave-dominated deltas, and barrier islands.

In addition to changing vertically, sedimentary rocks also change laterally. On either side of the tan, massive sandstone river channel deposit are dark gray, thin-bedded siltstones and shales. The thin, dark layer beneath the channel is coal. Note how the coal is thinner under the channel than on either side. The stream that deposited the channel probably eroded out some of the coal before depositing the channel sand.

The overall bedding is horizontal, but within some of the layers there is an internal, consistent layering at an angle. These cross beds were formed by the slip faces on migrating dunes. Cross bedding, Unaweep Canyon

These wave ripples, common in the tidal zone, form where currents oscillate back and forth, and no direction is dominant. Symmetrical ripple marks, Dinosaur Ridge

These ripples form where the current is consistent in one direction, shown by the pen point. Asymmetrical ripple marks, Dinosaur Ridge

In a geologic "ugly duckling becomes a swan" saga, sediments that were originally just gunk were later patiently sculpted by wind and water, pressed, and finally lifted to prominence as some of the state's most imposing landmarks. The sandstones of Colorado National Monument, the reddish-brown siltstones and mudstones of Owl Canyon, and the Flatirons that flank Boulder are all sedimentary rocks. Other sedimentary deposits include massive limestone formations around Leadville, the evaporites of the Eagle Valley, chalks of the eastern plains, coals near Trinidad, oil shale in western Colorado, and the thick shale of eastern Colorado.

Right: These beds are deposited in sudden pulses of sediment. The coarser sand settles out first, followed by the finer particles. Each bed grades from gray sand upward into darker shale. Graded bedding, highway 550 south of Ouray

Lower right: Sometimes microbiotic material grows on grains or fragments forming concentric rings similar to the layers in stationary stromatolites. Oncolites differ from stromatolites in that they were rolled around by periodic wave action. Oncolites, Pennsylvanian, near Molas Pass US-550

Below: These patterns form as conical shapes on a sandy bottom where small whirlpools scour out a deep, narrow upstream hole and then dissipate downstream. Flute casts, east of Colorado Springs

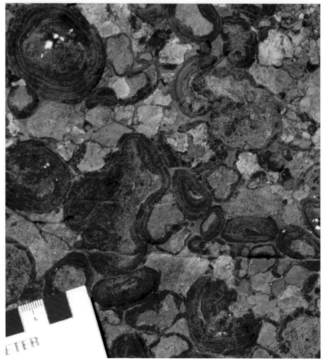

Sedimentary rocks differ from the crystalline igneous and metamorphic rocks in that most are comprised of fragments of pre-existing rocks. The names of Colorado's sedimentary rocks are largely derived from the size of the fragments of which they are composed. Conglomerates, with their gravel- and cobble-sized fragments, are the largest and most easily discernible. You can also see grains in sandstone with the naked eye, but you will need a magnifying glass to see the grains in siltstone, and a microscope to study shale and mudstone. The common carbonate rocks are limestone and dolomite, which are made up of grains of calcium carbonate (commonly fragments of calcareous fossils) and calcium-magnesium carbonate, respectively. Evaporites are composed of chemically precipitated minerals such as halite (salt) and gypsum.

In Colorado, sedimentary rocks can be any color of the rainbow. The most common, however, are gray, tan, brown, red, maroon, green, and black. The color is an important key to determining a rock's characteristics and history. To a trained eye, color reveals the minerals in the parent rock and shows the oxygen levels and water chemistry present in the environment during and after the material was deposited. It is also possible to determine how the rocks weathered and their attraction for lichens. As you drive along I-70 from Vail Pass to the Utah border, thick, colorful deposits of sedimentary rocks of all ages are on display—all you have to do is look out the window.

INTERPRETING ANCIENT SEDIMENTARY ENVIRONMENTS

How do geologists know that a particular sedimentary rock formed in a particular environment? Of course, they can never really know the real story. But, in 1795 a geologist developed a concept that is the next best thing to being there: "The present is the key to the past." The idea is that by studying the characteristics found in modern depositional environments and comparing them to similar features found in ancient rocks, one can solve the mystery.

This coarse-grained sedimentary rock makes up the resistant ledges above Castle Rock.

Raindrop imprints in Permian dune sandstones

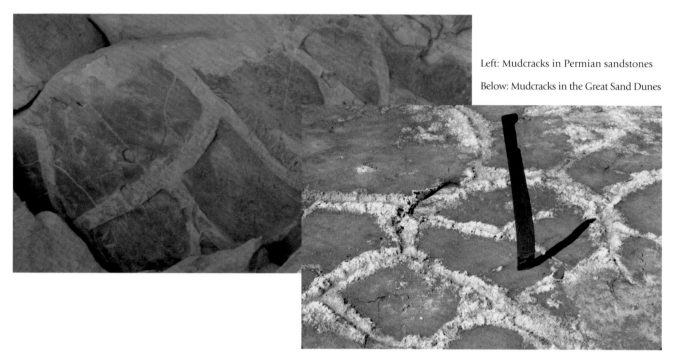

Left: Mudcracks in Permian sandstones

Below: Mudcracks in the Great Sand Dunes

Geologists, operating much like police detectives, piece together different sorts of clues in the rocks that provide circumstantial evidence of the type of environment in which the rocks were deposited. Some sedimentary structures can quickly rule out particular environments. For example, reptile footprints and raindrop imprints rule out a deep ocean environment.

In Colorado, we are fortunate because we can study the features of modern dunes in the Great Sand Dunes National Park and compare them to the ancient deposits of the 250 million-year-old Lyons sandstone, found along the eastern flank of the Front Range. More than ten different sedimentary features found in the present day Great Sand Dunes are also found in the ancient rocks quarried near Lyons. This gives the jury of the geologic community confidence in convicting the ancient rocks of having been deposited as dunes in a desert environment.

Reptile footprints in Permian dune sandstones

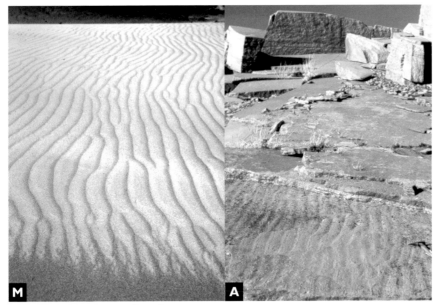

Transverse ripples on steep dune face

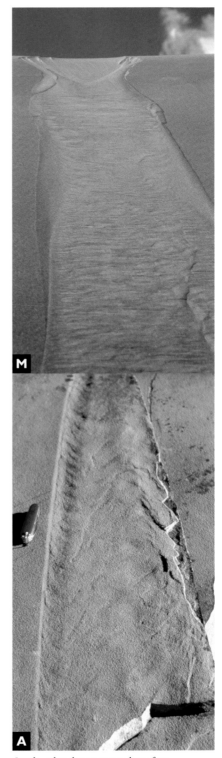

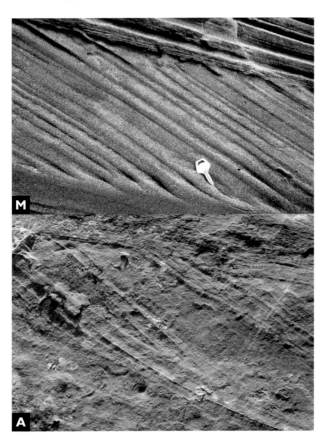

Cross laminations

Geologists commonly study the characteristics of modern sediments in order to interpret the origin of ancient sedimentary rocks. In each pair, the modern (M) example is from the Great Sand Dunes and the ancient (A) is from Colorado sandstone more than 200 million years old.

Sand avalanche on steep dune face

COLORADO'S FOSSILS: DENIZENS OF THE DEEP PAST

Colorado fossils, usually found in sedimentary rocks, have enriched museums around the world for nearly 150 years. Shortly after the Civil War, the USGS began explorations of the Colorado Territory. Their reports, published in the 1870s, are replete with gorgeous drawings of fossils of all types and ages.

One of their fascinating discoveries was a multitude of perfectly preserved insects near Florissant. Awed by the treasures surrounding him in 1873, one of the USGS geologists was inspired to write, "When the mountains are overthrown and the seas uplifted, the universe at Florissant flings itself against a gnat and preserves it."

Fossilized Insects

Florissant Fossil Beds National Monument is unique. It protects a site where ash from volcanic eruptions 34 million years ago trapped and preserved an entire ecosystem. Rangers conducting tours stress that while Florissant lacks the most glamorous fossils—dinosaurs—the insects and plant specimens provide more information about habitat and soft-bodied denizens than other sites. These fossil beds are internationally renowned for the variety and number of specimens yielded since their discovery.

Paleontologists have collected more than 60,000 specimens for museums and universities around the world. Fossils of more than 140 species of trees and other plants, including the world's first roses, were discovered at Florissant.

Above right: Fossil butterfly; Above left: Wasp;
Right: Maple leaf from Florissant Fossil Beds National Monument

Adult and child *Stegosaurus,* mounted dinosaur skeletons at the Denver Museum of Nature & Science

More than 1,100 species of insects are preserved as exquisitely detailed impressions in the shale. A delicate butterfly fossil clearly shows its antennae, legs, hairs, and wing pattern. Flies, too, are preserved, and in such detail that the lens in each eye can be counted. In stark contrast to the remains of tiny creatures, massive, petrified redwood stumps stud the site, ample proof that giants were also a part of this ancient world.

Fossilized, giant Sequoia stump at Florissant Fossil Beds National Monument

Dinosaurs

A unique period in Colorado's history began in 1870 when a single tail vertebra, the first dinosaur specimen found in western North America, was discovered in Middle Park. A year later, John Wesley Powell, the Grand Canyon explorer and first director of the USGS, remarked on the "reptilian remains" he observed in the area that was later to become Dinosaur National Monument.

Several years later, dinosaur bones were discovered near Cañon City. When word of this discovery spread, the state's reputation as a repository of world-class specimens solidified. Colorado was suddenly the metaphorical bulls-eye in a raucous scientific stampede that came to be called the "Great Dinosaur Wars." Eastern museums rushed teams of scientists to Colorado where a fierce battle was joined. Amid rumors and accusations of sabotage, spying, and claim-jumping, competitors scrambled to discover, name, and subsequently cart off the best and biggest skeletons. As a result of all this frenetic activity, dinosaur specimens from Colorado ended up in the museums of Pittsburgh, Cleveland, Chicago, Yale University, and the Smithsonian.

Colorado's dinosaur fossils cover a wide spectrum—coming from all three periods of the Mesozoic Era—and represent all six dinosaur groups and twelve of the fifteen known dinosaur families. Specimens recovered in the 1800s are still some of the best fossils of two of the groups of dinosaurs.

Colorado's State Fossil, *Stegosaurus*

Vertebrates

Later, fossils collected near Cañon City in 1887 again focused international attention on Colorado and its treasures. Strata there contained fossils of the world's oldest known vertebrates. Samples collected in Ordovician sandstones revealed at least three different types of fish that were 50 million years older than most of the fish fossils found around the world.

In addition to the fish, the outcrop in what is now the Indian Springs Trace Fossil Natural Area is unusually rich in markings or traces of animals that lived 450 million years ago. Study of twenty-five types of trace fossils—including tracks and burrow patterns of horseshoe crabs, brachiopods, and trilobites—give clues to the behavior of creatures living in the mudflats of an ancient tidal lagoon.

World's largest mounted skeleton of a dinosaur, a *Brachiosaurus* from Colorado who now resides on Concourse B of Chicago's O'Hare Airport

Allosaurus and *Stegosaurus*, a predator attacking the Colorado state fossil

Contemporary Fossil Discoveries

Despite 150 years of enthusiastic collecting in Colorado, it is impressive that important fossil discoveries continue to be made. In the past thirty years, scientists have unearthed the world's first articulated *Stegosaurus* skeleton; three of the world's four largest dinosaurs; the largest dinosaur trackway in North America; a huge palm forest; one of the world's most diverse leaf fossil sites; an eight-foot long mammoth tusk; and *Tyrannosaurus rex* bones.

The state's fossil riches are so extensive that paleontologists at the Denver Museum of Nature & Science have, with good reason, concluded that any excavation in metropolitan Denver will yield fossils. With Denver growing by leaps and bounds, reviewing each new hole is a formidable task.

Most of the recent finds have been fossils of plants and animals that occupied the Denver area during the past 67 million years. Excavations for the Denver International Airport uncovered hundreds of fossil palm fronds and twelve species of flowering plants. One of the palm fossils was six feet wide and eleven feet long. A trench in Littleton uncovered a mammoth's tusk about eight feet long weighing 500 pounds. A house foundation in a southern Denver suburb yielded *T. rex* bones, a particularly significant find since there are fewer than thirty *T. rex* skeletons in the world. During construction of Denver's new baseball stadium, paleontologists removed a dinosaur rib that was sticking up just behind what is now home plate.

Another amazing fossil site along I-25 near Castle Rock has yielded more than 115 broad-leafed flowering plants with leaves similar to those of modern tropical rainforests. The most complete fossil cycad (tropical, thick-stemmed, fern-like tree) ever found also came from this site. All of the species of modern cycads are found in rainforests, leading scientists to conclude that the climate of this area was once drastically different from the semi-arid present. In 2002, a six-foot-thick, 150-foot-high tree trunk was discovered that confirmed the rainforest interpretation. The find also clinched the interpretation that Earth's climate was much warmer 64.1 million years ago. The Colorado Piedmont had a climate similar to present-day Miami's, with 120 inches of rainfall and an average temperature of 75 degrees Fahrenheit. Although today's Rocky Mountains are spectacular, imagine how beautiful they must have appeared looming up out of a tropical rainforest.

Tropical plant from the Castle Rock rainforest site

From a rainforest to a sea, Colorado had more than one strikingly different environment. Excavations in the Pierre Shale closer to the mountains, where rock layers are tilted upwards, unearthed fossils from 68 to 90 million years ago when Colorado was at the bottom of a several-hundred-foot-deep sea. Fossils from this shale include giant clams, mosasaurs, plesiosaurs, fish, sharks, oysters, crabs, coiled squid relatives known as ammonites, and flightless penguin-like birds.

Common fossils found along the eastern flank of the Front Range in Pierre Shale

Artist interpretation of the Denver area shortly after the beginning of the Tertiary (64 million years ago) Period, when life had begun to recover from the asteroid impact. All plants are based on fossils actually found in the area.

Notable Fossil Rock Units

Three Colorado rock units are world famous for their treasure troves of fossils: the White River Formation in northeastern Colorado, the Green River Formation in the northwestern part of the state, and the Morrison Formation found in numerous locations around the state. Other units also have rich fossil assemblages, but these three are extraordinary.

Fossils of camels, elephants, horses, mammoths, hippos, and rhinoceroses were entombed between 34 and 37 million years ago, and are excavated regularly on the eastern plains out of the White River Formation, one of the richest fossil mammal beds in the world. The bones reveal that a cross between a pig and sheep was the most common mammal in Colorado at that time. Predecessors to the camel and antelope were also common. Large "thunder beasts" with two horns on their snouts were the largest mammals roaming the plains, larger even than the true rhinoceroses that were in Colorado by then.

The Denver Museum of Nature & Science has first-class fossil exhibits. Other displays are located at the Morrison Museum of Natural History, University of Colorado Museum in Boulder, Delta County Museum in Delta, Dinosaur Depot in Cañon City, and the Museum of Western Colorado in Grand Junction.

Several outdoor sites around the state provide exceptional opportunities to view fossils intact and are well worth a visit. Among them are Riggs Hill in Grand Junction and Dinosaur Ridge west of Denver, which offers guided tours. Rabbit Valley west of Grand Junction has a twenty-foot long neck of a Sauropod (gigantic, four-footed, plant-eating dinosaur) in place, and Dinosaur National Monument on the Colorado–Utah border has preserved a dinosaur quarry that makes for a fascinating visit amid spectacular scenery.

Fossilized neck vertebrae of a juvenile *Diplodocus* along the "Trail of Time" in Rabbit Valley

In the pages of strata in northwestern Colorado, you can find early Cenozoic fossils marking the beginning of the "Age of Mammals." The famous Green River Formation contains beautiful fossils of fish, scorpions, beetles, frogs, hundreds of insect species, and more than 100 species of trees, including palms that lived about 50 million years ago. Foreign and local collectors flock to the Bureau of Land Management areas near Douglas Pass to find fossils of the Green River Formation. Cenozoic strata around the state often yield fossils of petrified wood. One building in Lamar used forty tons of petrified wood in its construction.

The 150 million-year-old Morrison Formation of Colorado is a spectacular source of fossilized remains of Jurassic dinosaurs that roamed a vast lowland. The several-hundred-foot-thick layers of the Morrison Formation yield dinosaur bones and tracks including the world's largest and smallest dinosaurs. These range from the eighty-foot-long, plant-eating quadruped Brontosaurus (*Apatosaurus*) to othnielia, a fleet-footed, chicken-sized, bipedal dinosaur.

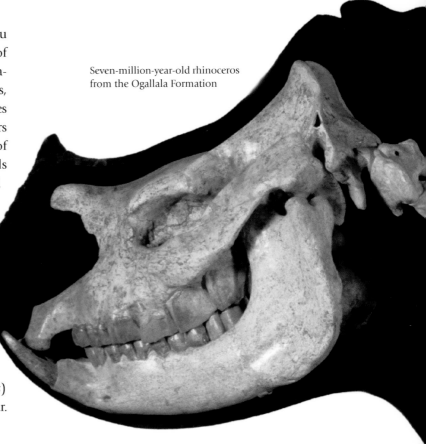

Seven-million-year-old rhinoceros from the Ogallala Formation

Forty-eight-million-year-old beetle from the Green River Formation

In 1932, a building in Lamar was constructed with forty tons of petrified wood. It even has a petrified log floor and is now part of Stagner Tire Store.

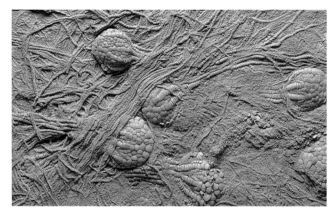

Eighty-million-year-old floating crinoids from the Mancos Shale

Since the first fossil dinosaur finds in the Morrison Formation of Colorado in 1877, more than a dozen major dinosaur quarry sites have been opened and worked, along with scores of smaller digs. The rocks yielded an enormous amount of dinosaur remains for museums and universities. The complete *Stegosaurus* on display at the Denver Museum of Nature & Science is one of only six complete skeletons in the world and was collected from the Morrison Formation in 1937. In Mesa County, scientists probing the Morrison Formation found three of the four largest dinosaur skeletons in the world, and a nest of dinosaur eggs was found near Cañon City.

Colorado also has beautifully preserved dinosaur footprints. A pavement-like plane of Morrison limestone located in a remote canyon of the Purgatoire River in southeastern Colorado was referred to as "the elephant crossing" as early as 1920. Large prints at this site received little subsequent attention until the early 1980s when it was determined to be the longest continuous trackway in North America with at least 1,300 dinosaur tracks. Some areas are so trampled that individual footprints are hard to discern. These footprints tell us things about the dinosaurs, how they moved and behaved, for example, that bones alone cannot.

In addition to its justly famous dinosaur finds, the Morrison Formation has fossils of turtles, pterosaurs (winged dinosaurs), crocodiles, lizards, mouse-sized mammals, frogs, fish, clams, snails, crustaceans, and plants. The

fossil richness of the Mesozoic strata is not restricted to land creatures. Of the many fossil "seashells" found in the marine rocks of the Cretaceous Interior Seaway, researchers find the ammonites particularly interesting because they evolved so rapidly during the Cretaceous Period. Some of these coiled ammonites near Kremmling are two-feet across. Associated with these giant ammonites is an assemblage of more than sixty species of rare tropical and subtropical marine fossils.

The information about the history of life on Earth that can be obtained from fossils is irreplaceable, so a word of caution is warranted. Inexpert collecting, or failure to maintain precise information on the original location, rock type, or other conditions of a fossil occurrence, can damage fossils or cause them to lose their context and much of their value as objects of study. This is why collecting fossils on federal lands is regulated by federal agencies. It is important to check regulations and obtain necessary permits (on state or federal lands) or permission (on private lands) if you are interested in collecting fossils.

Tracks of an adult and child dinosaur at Dinosaur Ridge

Check out this illustrated road map to the fossil stars of Colorado.

Giant Cretaceous ammonites near Kremmling

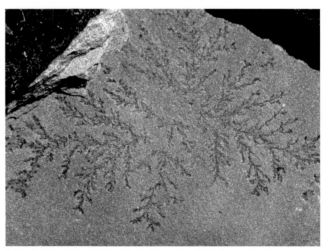

It looks like a fossil, but it isn't. These fern-like shapes fool many people. They are actually inorganic deposits of the mineral Pyrolusite MnO_2. The dendrites are precipitated from groundwater in the cracks of this Permian sandstone.

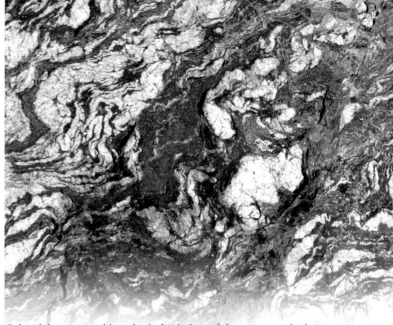

The Big Change: Colorado's Metamorphic Rocks

As the name indicates (meta = change, morph = form), metamorphic rocks are pre-existing igneous or sedimentary rocks that have been altered, or metamorphosed, deep within Earth's crust. The rocks changed form in response to intense fluctuations in temperature, pressure, shearing stress, or chemical environment. A cross section of Colorado's metamorphic rocks can be seen in road cuts along much of I-70 from Denver to Copper Mountain.

Colorado's geologic history includes mountain building events, many of which were accompanied by the intrusion of igneous bodies, and an increase in the temperature of the whole region. During these periods, two types of metamorphism, contact and regional, took place, occasionally at the same time.

Contact metamorphism occurs when hot magma intrudes into cooler rock. The intrusion heats the surrounding rock making the low-temperature minerals unstable. These minerals change to minerals that are stable at the new, higher temperatures.

Colorado's metamorphic rocks rival paintings of abstract expressionists. Red splotches of lichen dot the swirling layers of these high-grade metamorphic rocks near Georgetown.

Regional metamorphism often accompanies massive intrusions of igneous rock across large areas. This is the process that created the core of many Colorado mountain ranges. During regional metamorphism, as the granitic magma intrudes and batholiths form, the temperature of the pre-existing rock increases and its minerals adjust by changing to forms stable at higher temperatures and pressures. If the temperatures and pressures become high enough, the rocks can even partially melt.

Folia of mica

Thin section of metamorphic gneiss showing the parallel bands of mica-forming folia. Notice how the minerals are intergrown, rather than forming rounded grains as in sandstones.

Contact metamorphism of the Leadville limestone by intrusion of a Tertiary stock created the famous Yule Marble. The marble is exceptionally pure in color and grain size, and is a standard for conducting laboratory experiments on rock deformation.

MESSAGES IN STONE

Yule Marble is quarried in huge rooms inside the mountain.

This massive block of Yule Marble weighed a quarter of a million pounds and was the largest single block of stone quarried in Colorado, until a slightly larger one was quarried in 2002.

Belts of high-temperature metamorphic rock form nearest the batholith; medium- and low-temperature metamorphic rocks form in cooler areas progressively farther away from the batholith. Index minerals characterize each of these grades of metamorphic rocks. Chlorite and biotite indicate low-grade metamorphism (400 to 750 degrees Fahrenheit and twenty-one tons of pressure per square inch). Garnet and staurolite are products of medium-grade metamorphism, and kyanite and sillimanite characterize high-grade metamorphism (950 to 1,475 degrees Fahrenheit and 110 tons of pressure per square inch).

In the course of a drive through Big Thompson Canyon from Loveland to Estes Park, you traverse all of the mineral zones of regional metamorphism. The presence of biotite near the mouth of the canyon signals an area of low-grade metamorphic rocks. Farther up the canyon are the garnet and staurolite zones indicative of formations subjected to higher temperatures. Within five miles, you reach the community of Drake in the highest (sillimanite) temperature and pressure zone of regional metamorphism. Exposed in the hills above Drake are coarse-grained pegmatites that are vein-like offshoots of the batholith. The high-grade rocks and pegmatites are indicators that you are approaching the 600-square-mile granitic batholith that surrounds Estes Park.

Yule Marble's uniform, pure-white color led to its use in The Tomb of the Unknowns in Arlington National Cemetery.

Yule Marble was used in the Lincoln Memorial.

Around Estes Park and Rocky Mountain National Park, the metamorphic rocks were raised to temperatures and pressures at or near their melting point. This gave rise to migmatites, an intimate mixture of igneous and metamorphic rocks ("migma" means mixed igneous and metamorphic material). Migmatites are found throughout much of

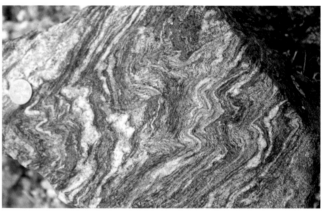

The many faces of Colorado migmatites, fascinating displays of mixed igneous and metamorphic rock

the Front Range and other areas in Colorado. Geologists are not sure whether the igneous part of these migmatites formed because these metamorphic rocks got so hot that they partially melted or whether the rocks were so close to the batholith that they were just extensively intruded along many layers.

The dominant metamorphic rock types in Colorado are gneiss, schist, amphibolite, and quartzite. In low-grade metamorphic rocks, geologists often can find evidence of the parent material. In high-grade metamorphic rocks, the original rock type is much more difficult to discern, but occasionally it is still possible.

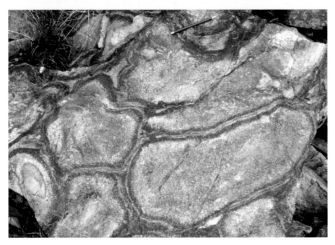

Examples of original structures that survived Precambrian regional metamorphism in Colorado: Upper left: Cross bedding highlighted by white lines in 1.8-billion-year-old metasandstone in the Needle Mountains; Upper right: Aligned pebbles in metaconglomerate older than 1.4 but younger than 1.7 billion years; Lower left: Pillow structures in 1.8-billion-year-old metabasalt in the Needle Mountains; Right center: Tuff breccia fragments in 1.8-billion-year-old metavolcanic rock in the Needle Mountains; Lower right: Pebbles in metaconglomerate in Coal Creek Canyon.

Contortions in the Rocks: Colorado's Unconformities, Faults, and Folds

Mountain building episodes throughout Colorado's history created geologic structures that are beautifully exposed across the state. Ancient structures that formed more than a billion years ago can be seen today and attest to the longevity of mountain building activity in Colorado. Young faults that offset 10,000-year-old glacial deposits and present-day earthquakes indicate the continuing rise of the mountains.

When rocks respond to the forces acting in Earth's crust, they bend and/or break. When strata bend, the resulting structure is a fold. When strata break, a joint or fracture is formed. If the rocks move after they break, the fracture becomes a fault. In the faults, folds, and unconformities of Colorado we can look back to see how Earth continually makes and remakes itself. Colorado has every kind of fold and fault known to geologists.

UNCONFORMITIES: GAPS IN THE RECORD

Sometimes missing rocks provide important information. Unconformity is the general term for missing pages of Earth history. There are three types of unconformities: angular unconformity, nonconformity, and disconformity.

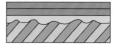 Angular unconformity is an erosional surface separating steeply dipping rock layers below from gently dipping layers above.

 Nonconformity is an erosional surface separating igneous or metamorphic rocks below from sedimentary strata above.

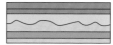 Disconformity is an erosional surface separating horizontal strata below from horizontal strata above and where there is a gap in time.

Angular unconformity at Box Canyon in Ouray

Five-hundred-million-year-old sand-stones rest on 1.7 billion-year-old igneous and metamorphic rocks in Glenwood Canyon

On the right are 300 million-year-old sedimentary rocks that rest non-conformably on 1,700 million-year-old metamorphic rocks on the scrub-covered slopes. The contact is exposed and labeled with a bronze plaque in the upper parking lot of Red Rocks Park & Amphitheater.

Exposed in the parking lot of Red Rocks Park & Amphitheater near Denver are red, 300-million-year-old sedimentary rocks resting on gray 1,700-million-year-old metamorphic rocks. About 1.4 billion years of history are missing in the geologic record. This gap is a nonconformity.

When scientists study unconformities, they try to answer two questions: (1) Were the missing rocks deposited and later eroded, or were they not deposited?, and (2) What geological events (e.g., mountain building, uplift, intrusion, metamorphism) occurred during the time gap? Questions raised by the absence of strata at Red Rocks are at least partially answered by rocks in other areas of the Front Range. Studying these areas, geologists discovered that the missing strata were indeed deposited at Red Rocks and then later removed by erosion during an uplift asso-ciated with mountain building.

Nonconformities that occur between sedimentary and older crystalline rocks, such as at Red Rocks, are also exposed in other places in Colorado. There are three particularly striking examples. In Glenwood Canyon, sandstone over-lies metamorphic and igneous rocks with 1.2 billion years of history missing at the nonconformity. In the Black Canyon of the Gunnison and Colorado National Monu-ments, sedimentary rocks rest nonconformably on igneous and metamorphic rocks with a gap of 1.5 billion years. Actually, if we could see beneath the sedimentary rocks at Earth's surface, we would find that there is a noncon-formable surface at the top of the igneous and metamor-phic rocks everywhere in Colorado.

Rocks and Contortions

Angular unconformity: Conglomerate overlies dipping shale in Little Snake Canyon. About 150 million years of Earth history occurred between the deposition of the rocks below and those above the unconformity.

Angular unconformities are the most spectacular of all the unconformities and reveal a tremendous amount of history. An angular unconformity consists of older, underlying rocks dipping at a steeper angle than the younger, overlying strata. The angular unconformity in Box Canyon near Ouray is so striking that it often appears in geology texts. This beautiful unconformity reveals a history that is applicable to most other angular unconformities. The steeply dipping Precambrian strata were originally deposited as horizontal layers and then buried. Later, the layers tilted and folded to a vertical position during a mountain building event and were eventually eroded and truncated. When the vertical Precambrian strata subsided below sea level, the Devonian marine sandstone was deposited on top of them. These strata were then lifted to their present elevation of 8,000 feet where they continue to erode. That's a lot of information from one outcrop.

Angular unconformity: Volcanic rocks overlie sedimentary rocks near State Bridge. About 270 million years of Earth history occurred between the deposition of the rocks below and those above the unconformity.

Left: Limestone disconformably overlying sandstone in Canyon of the Lodore. The sequence of layers look continuously vertical, yet the fossils tell us that about 110 million years of history is missing.

Sandstone disconformably overlies 1-billion-year-old sedimentary rocks in Canyon of the Lodore, Dinosaur National Park. The rocks look as though there could have been continuous deposition, yet about half a billion years of history is missing across the disconformity.

JOINTS AND FAULTS

When rocks are stressed, they often respond by breaking into fractures called joints. Commonly, joints occur in sets in which numerous joints are parallel to each other. Joints are abundant throughout the rocks of Colorado.

Basically, a fault is a break in Earth's crust where the broken blocks of rock move with respect to one another. This movement normally generates earthquakes. The amount of movement is the fault displacement.

Left: Three joint sets break the sandstone at Dinosaur Ridge into very consistent patterns

Joint sets enhanced by spheroidal weathering along the Purgatoire River

The next step is to determine whether the hanging wall has moved up or down relative to the footwall. In normal faults, the hanging wall moves down relative to the footwall. In a reverse fault, the hanging wall moves up relative to the footwall. A thrust fault is a special case of a reverse fault where the dip of the fault plane is thirty degrees or less.

Some faults in Colorado have displaced the crust an astounding amount. For instance, the Sangre de Cristo fault displaces the Sangre de Cristo Mountains nearly four miles

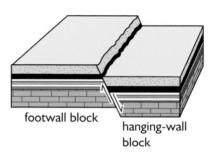

Normal fault

It will come as a surprise to many readers that Colorado has thousands of faults throughout the state that range in displacement from inches to miles. You name the type of fault and chances are Colorado has it—young faults, old faults, small faults, huge faults, ductile faults, normal faults, reverse faults, thrust faults, and possibly strike-slip faults (although this last is quite controversial).

The first step in classifying a fault is to determine which block is the footwall and which is the hanging wall. Miners originally came up with these terms because ore deposits are often found along faults. Since faults are breaks in Earth's crust, it is natural that ore-forming fluids would use them as passageways as they move upward through the crust and eventually precipitate out into an ore deposit. When a miner digs a hole along the ore deposit (fault), they are walking on the footwall, with the hanging wall overhead.

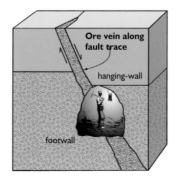

CGS geologist sitting on footwall block; hanging wall has moved down about fifteen feet in this fault in Tertiary sedimentary rocks west of Trinidad, a normal fault.

vertically from the same rocks in the San Luis Valley. This displacement did not occur in one giant earthquake, rather it is the result of movements of five to twenty feet at a time. Studies of this fault show that it is still active and can be expected to move again in the future. As the range continues to rise, large earthquakes loom as a possibility in the future. Other young, normal faults are found along the eastern flanks of the Sawatch Range and the Gore Range.

An excellent, large-scale example of a thrust fault is the Williams Range fault bounding the west side of the Front Range exposed along I-70 near Dillon. A tunnel bored through the fault documents that Precambrian metamorphic rocks moved up and over Cretaceous rocks horizontally at least a third of a mile.

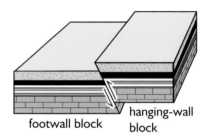

Reverse fault

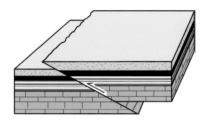

Thrust fault

Hanging wall has moved up about twenty feet in this fault in Gore Canyon. The fault displaces Precambrian rocks.

Hanging wall has moved up in this fault east of Silverthorne on I-70. Note how the thrust fault is overlain by a fold.

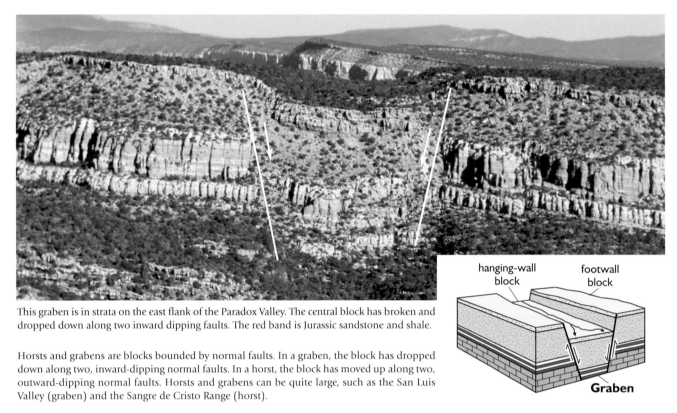

hanging-wall
block

footwall
block

Graben

This graben is in strata on the east flank of the Paradox Valley. The central block has broken and dropped down along two inward dipping faults. The red band is Jurassic sandstone and shale.

Horsts and grabens are blocks bounded by normal faults. In a graben, the block has dropped down along two, inward-dipping normal faults. In a horst, the block has moved up along two, outward-dipping normal faults. Horsts and grabens can be quite large, such as the San Luis Valley (graben) and the Sangre de Cristo Range (horst).

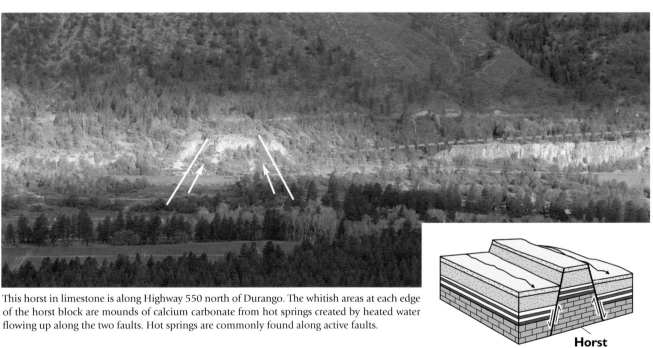

Horst

This horst in limestone is along Highway 550 north of Durango. The whitish areas at each edge of the horst block are mounds of calcium carbonate from hot springs created by heated water flowing up along the two faults. Hot springs are commonly found along active faults.

FOLDS

The folds in Colorado's rocks, those sinuous bands of seemingly liquid color preserved in a flat, cold face of stone, are intriguing to geologists and lay viewers alike. For geologists who have studied Colorado rocks, the variety is surprising. The state has metamorphic folds, basement-cored folds, salt-cored folds, monoclines, syndepositional folds, anticlines, synclines, domes, basins, refolded folds, evaporite-flowage folds, collapse folds, disharmonic folds, and forced folds.

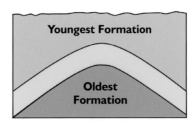

Anticline: folds with limbs that dip outward away from the hinge of the bend; convex upward

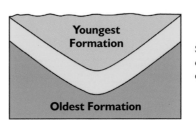

Syncline: folds with limbs that dip inward toward the hinge of the bend; concave upward

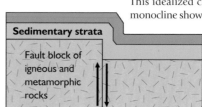

Monoclines

Monoclines are folds that, in contrast to anticlines and synclines with two dipping limbs, have only one limb with horizontal beds on either side of the steep limb. Various places in Colorado show the relationship of monoclines in the sedimentary strata to block faulting in the underlying Precambrian crystalline rocks. It is the differential movement and shape of these underlying fault blocks of igneous and metamorphic rock that ultimately determines the shape of the overlying monoclines. Monoclines range in size from rather small, north of Fort Collins, to huge monoclines associated with the Uncompahgre Plateau and White River Plateau basement blocks.

Well exposed monocline on the northwest end of the Uncompahgre Plateau

This idealized cross section of a Colorado monocline shows igneous and metamorphic rocks that break into blocks, which move relative to each other. The overlying sedimentary strata fold over the edges of the blocks forming a monocline.

Sedimentary strata

Fault block of igneous and metamorphic rocks

Double monocline: aerial view of two monoclines northwest of Colorado National Monument. The strata flow over the edges of basement blocks like a two-tiered waterfall

Horizontal layers

Steeply dipping layers

You can see a readily accessible monocline east of Livermore on Larimer County Road 80.

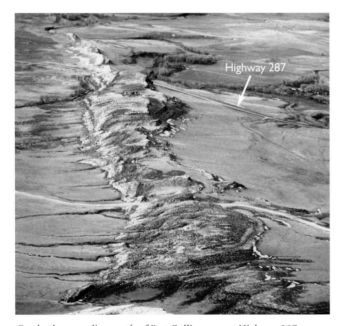

Highway 287

Grayback monocline north of Fort Collins crosses Highway 287

Steeply dipping layers

Large monoclines abound in Dinosaur National Monument. Notice the horizontal layers in the foreground and in the cliff top. Steeply dipping beds of the monocline are in the slopes of the cliff.

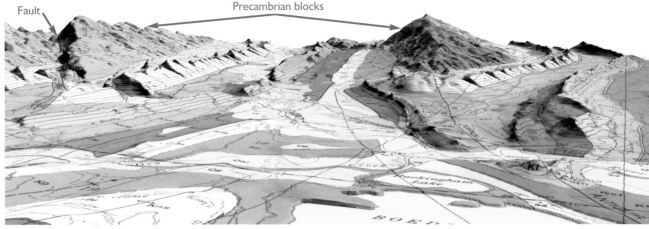

Basement-cored anticlines west of Loveland: the fold on the right is so spectacular it is found in geology books dating back to 1873. In this area you can see that the Precambrian basement rock is broken into differentially uplifted and eastward-tilted blocks, and that the geometry of the folded sedimentary strata is a direct reflection of the shape of the basement blocks.

Basement-Cored Folds

During the Laramide mountain building event, igneous and metamorphic rocks broke into large blocks throughout Colorado. These broken blocks pushed up into the overlying sedimentary rocks forcing them to fold over the edges of the Precambrian blocks. The shape of the block determined the shape of the folds in the overlying sedimentary strata. Where the blocks moved up without any rotation, the overlying sedimentary rocks formed monoclines.

Where blocks rotated as they were differentially displaced, they formed asymmetrical anticlines and synclines in the overlying sedimentary rocks. These anticlines are characterized by one limb dipping gently (usually twenty degrees or less) and the other limb(s) dipping steeply (commonly greater than fifty degrees and often vertical to overturned). The gentle limb is the one resting passively on the top of the rotated Precambrian rocks. The steep limb is the one folded over the edge of the rotated block. These basement-cored structures are important traps for petroleum throughout Colorado and Wyoming. One of the structures has yielded nearly a billion barrels of oil at Rangely Oil Field in northwest Colorado.

Excellent examples of Precambrian-cored folds are found along the mountain front in a belt from Lyons to north of Fort Collins. This small area is particularly important because of the variety of fold shapes and the excellent exposures of the blocks of Precambrian rocks that allow geologists to study the relationship of the folds to the underlying blocks in their cores.

This anticline, four miles east of Lyons on Rabbit Mountain, is a particularly nice example of a fold reflecting the geometry of the basement block that moved up into the overlying sedimentary strata and created a fold in those strata. It has a planar limb gently dipping to the left, and a steep nearly vertical limb on the right.

MESSAGES IN STONE

Salt Anticlines

Long, linear anticlines with cores of salt are found in the Eagle and Carbondale areas of central Colorado and in the Paradox Basin of western Colorado. Thick accumulations of salt and other evaporites were deposited in the Paradox Basin during the Pennsylvanian Period. Salt was able to flow, and as younger deposits were being deposited on top of the salt, it began to move upward, folding the overlying sedimentary rocks into anticlines. Salt continued flowing upward into the cores of the anticlines for millions of years, creating angular unconformities in the sediments being deposited on the flanks of the growing folds. A well in the Paradox Valley penetrated 14,000 feet of salt in the core of this anticline.

A final chapter in the growth of these salt anticlines puts an intriguing twist on their geometry. Eventually, the salt reached the surface and eroded. It dissolved much more quickly than the overlying rocks and formed large, linear valleys over the crest of the anticlines. This rapid dissolution of the salt cores caused the rocks near them to collapse, creating the inward-dipping layers that define a syncline. The result is a syncline sitting on top of an anticline.

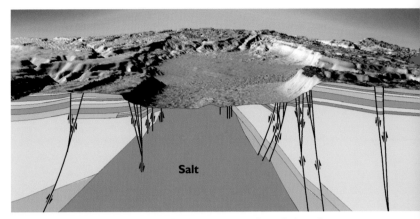

Paradox Valley salt anticline: the salt has been rising up through the sedimentary strata for hundreds of millions of years. The gray and blue layers are Pennsylvanian strata and the green layers are Mesozoic strata. At the far end of the valley is the collapsed syncline.

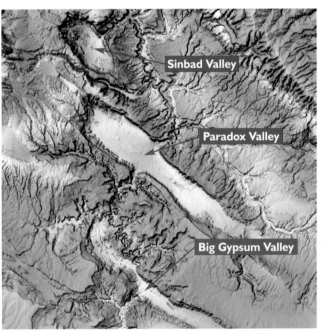

Paradox Basin salt anticlines: these northwest-trending valleys are geomorphic expressions of salt anticlines.

Complex small-scale folds are created when evaporites flow on a large scale.

In the Eagle and Carbondale areas, the surface collapsed as much as one-half mile vertically over huge regions because groundwater dissolved the underlying evaporites. We can reconstruct the history of the collapse by dating basalt flows, some of which are affected by the collapse and some of which are not affected.

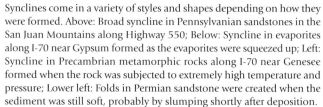

Synclines come in a variety of styles and shapes depending on how they were formed. Above: Broad syncline in Pennsylvanian sandstones in the San Juan Mountains along Highway 550; Below: Syncline in evaporites along I-70 near Gypsum formed as the evaporites were squeezed up; Left: Syncline in Precambrian metamorphic rocks along I-70 near Genesee formed when the rock was subjected to extremely high temperature and pressure; Lower left: Folds in Permian sandstone were created when the sediment was still soft, probably by slumping shortly after deposition.

To close-up "A" below

B

Anticline and synclines in metamorphic rocks near Blackhawk. Within these large folds are many small, tight folds that were formed during an earlier period of folding and were refolded by the second period.

Refolded Folds

The oldest folds in Colorado are in Precambrian metamorphic rocks. The layers were folded during different regional metamorphic events. When previously folded rocks were again subjected to heat and pressure, they were refolded and became refolded folds.

Early geologists studying the Precambrian structures of the Front Range found many clues indicating two periods of folding, but were unable to find a place where they could see both sets of folds in the same outcrop. Finally, they found an outcrop in Clear Creek Canyon east of Blackhawk that displayed both sets of folds.

It is fairly easy to pick out the large anticline and synclines. However, it requires careful searching to discern the smaller tight folds that were folded once during an earlier period of folding, then folded again and tightened by a second period of folding.

Close-ups of the tight, refolded folds in the outcrop near Blackhawk

Getting It On Paper: Geologic Maps, Cross Sections, and Formations

A geologic map depicts the aerial distribution of various rock types of different ages. Rocks are broken out into mappable units or formations. A formation is a rock type that is distinct enough to separate from the other rocks in the area, and thick enough to show accurately on the map. Each formation is represented by a unique symbol and color on the geologic map. The symbol indicates the rock's age and formation name. The geologic pattern and symbol for each formation are overprinted on a topographic map.

The geologic pattern (dark green) and symbol (Jwg), for the Jurassic Wingate sandstone, indicate the age and rock type and is overprinted on the topographic contour lines.

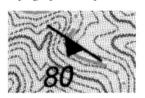

This symbol shows that the metamorphic layering at this spot is trending northwest southeast and is inclined into the ground at eighty degrees toward the southwest.

The first letter of the map symbol represents the formation's age (e.g., "Jwg" indicates that the formation is Jurassic in age). The geographic name that is the first part of the formation name refers to the place where it was first described and formally named. The Wingate sandstone was first described and designated at Fort Wingate, New Mexico. If you want to compare the rocks you are studying to the Wingate sandstone, then you should be able to go to Fort Wingate and see the rocks that were originally named and designated the Wingate sandstone. Sometimes, after more information is discovered, another place may be designated as a better place to reference the formation, but the original name sticks. This is true for the Wingate sandstone.

Some other well-known formations in Colorado and their original locations are: Morrison Formation, Morrison, Colorado; Pierre Shale, Pierre, South Dakota; Leadville Limestone, Leadville, Colorado; Castle Rock Conglomerate, Castle Rock, Colorado; Mancos Shale, Mancos, Colorado; and Fountain Formation, Fountain, Colorado.

Strata are usually deposited in nearly planar, horizontal layers—the Law of Original Horizontality. If the layers have been disturbed since deposition, they may now be at an angle other than horizontal. Indeed, they may even be vertical or overturned. The inclination of the layered strata are shown by symbols on a geologic map that depict the direction and amount of inclination.

Cross sections are drawn for the map to help the user visualize the third dimension of the geology. A geological cross section attempts to depict what you would see if you took a giant knife and cut vertically into the rocks and then turned it up on end to see the buried layers.

Geologic maps are useful in several ways. For instance, if you were going to visit a certain place, you could examine the geologic map of that area and find out what kind of rock is present. Also, suppose that you were interested in shale. You could determine from the map legend which formations contain shale. Then, you could peruse the map looking for those colors and know where to go in your search for shale. Or perhaps you want to look for dinosaur bones. You could go to the geologic map and see where the Morrison Formation outcrops and then head for those likely spots. (Remember not to collect any bones, if you want to stay out of trouble.)

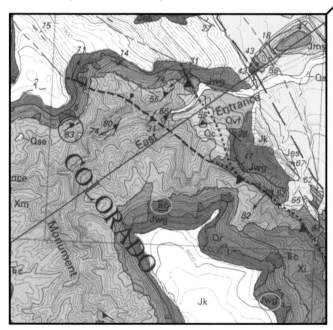

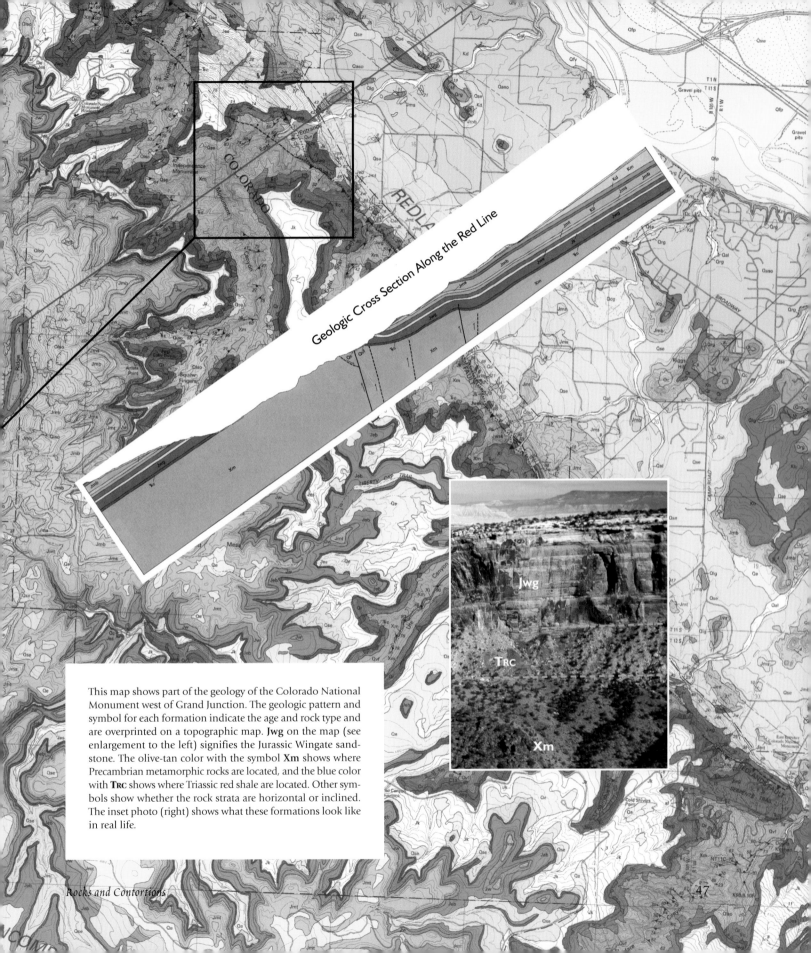

Geologic Cross Section Along the Red Line

This map shows part of the geology of the Colorado National Monument west of Grand Junction. The geologic pattern and symbol for each formation indicate the age and rock type and are overprinted on a topographic map. **Jwg** on the map (see enlargement to the left) signifies the Jurassic Wingate sandstone. The olive-tan color with the symbol **Xm** shows where Precambrian metamorphic rocks are located, and the blue color with **T**RC shows where Triassic red shale are located. Other symbols show whether the rock strata are horizontal or inclined. The inset photo (right) shows what these formations look like in real life.

Precambrian Rocks

This image shows the geologic map of the Colorado National Monument geologic map draped over a digital elevation model of the area. The combination of the two allows geologists to view the geology and topography from a variety of angles. The photograph shows what the actual rocks look like in the field. Both the photograph and the image show the faulted edge of the block of Precambrian rocks extending under an un-faulted monocline (outlined in dashed yellow) in the Jurassic Wingate sandstone.

GENERALIZED GEOLOGY OF COLORADO

Explanation

Unconsolidated Deposits of Quaternary Age

Sedimentary Rocks of Tertiary Age

Igneous Rocks of Cenozoic Age

Sedimentary Rocks of Mesozoic Age

Sedimentary Rocks of Paleozoic Age

Sedimentary, Metamorphic, and Igneous Rocks of Precambrian Age

The Story in the Rocks: Brief Geologic History of Colorado

Eons and Eons: Geologic Time

Scientists organize Earth's grand history and subdivide it into units of manageable time called eons, eras, and periods. In the early days of geologic study, we could only work out the relative ages of rocks—B is older than A and younger than C. Using just these relative ages, geologists constructed the Geologic Time Scale.

With the discovery of radioactivity near the turn of the twentieth century, geologists were able to assign ages to rocks in thousands, millions, and billions of years, thus adding absolute ages to the time scale.

The oldest dated rocks on Earth are about 4 billion years old. The oldest dated rocks in Colorado are about 2.7 billion years old and lie in a small area in the very northwest corner of the state.

Predictably, geologists can see more in the youngest rocks because these strata are the least disturbed, have endured the shortest history, and commonly contain more fossils than in the oldest rocks. This more detailed description and finer differentiation between strata explains the rather lopsided split between the three grand divisions of geologic time called eons. The Precambrian, which is the oldest portion of geologic time, constitutes the majority of Earth's history, but is split into only two eons, Archean and Proterozoic. By contrast, the Phanerozoic Eon is divided into three eras (Paleozoic, Mesozoic, and Cenozoic) and twelve periods.

These Colorado rocks have endured at least 2 billion years of Earth history and are now being broken up by the relentless processes of weathering. Hammered into bits and pieces, they will soon be transported by the forces of erosion to some distant shore. There they will be unceremoniously dumped as sand and mud.

COLORADO'S GEOLOGIC TIME SCALE

Era	Period	(Millions of years ago)	Major Geologic Events in Colorado
CENOZOIC	Quaternary	Present–1.8	Development of present topography, ice ages, huge dune fields, widespread mammalian extinction, widespread faulting, basaltic volcanoes, development of caves
	Tertiary	1.8–65	Rio Grande rifting and regional uplift by block faulting (to present elevations); major canyon cutting; catastrophic volcanoes erupting; erosion and basins developing; Laramide mountain building; igneous plutons intruding and mineralization along the Colorado Mineral Belt
	K/T Boundary	65	Asteroid impact causes worldwide extinction of plants and animals. End of the "Age of Dinosaurs," beginning of "Age of Mammals"
MESOZOIC	Cretaceous	65–144	Subtropical to tropical climate; shifting shorelines of Western Interior Seaway; deep- and shallow-marine and non-marine conditions; onset of Laramide mountain building; final retreat of marine waters; dinosaurs
	Jurassic	144–206	Lakes, swamps, braided and meandering streams; repeated invasion and withdrawal of sea; coastal dune deposits; dinosaurs
	Triassic	206–251	Semi-arid and arid conditions; mudflats, alluvial plains, dune fields adjacent to eroding Ancestral Rockies; deposition of "redbeds;" dinosaurs dominant life forms
PALEOZOIC	Permian	251–290	Dune fields; continued erosion of Ancestral Rockies; deposition of coarse "redbeds;" mass extinction
	Pennsylvanian	290–323	Widespread karst development on Mississippian limestones; shallow seas and evaporite basins; rise of the Ancestral Rockies; erosion of great volumes of rock and deposition into alluvial fans
	Mississippian	323–354	Widespread shallow sea
	Devonian	354–417	Widespread shallow sea; intrusion of kimberlite diamond pipes
	Silurian	417–443	Widespread shallow sea; uplift and erosion
	Ordovician	443–490	Local deepening of seas and cyclic sea level rise and fall; first vertebrates in the world
	Cambrian	490–545	Invasion of sea from west and east; dike intrusion; rapid development of hard-bodied life forms
PRECAMBRIAN PROTEROZOIC EON		545–2,500	Rifting, a major continent/island-arc collision; three periods of regional metamorphism; three periods of folding; formation of major shear zones; three periods of granitic intrusions; a period of major dike intrusion; deposition of a thick sequence of continental sedimentary rocks
PRECAMBRIAN ARCHEAN EON		2,500–4,500?	Granitic intrusion, regional metamorphism, and folding; metamorphism at 2.7 billion years ago

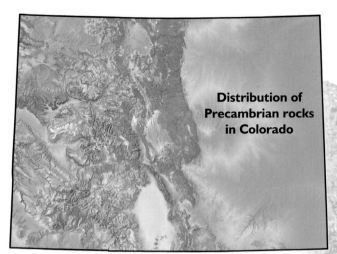

Distribution of Precambrian rocks in Colorado

The gray and white Archean metamorphic rocks in the lower two-thirds of the photo are the Owiyukuts Complex and are 2.7 billion years old, the oldest rocks in Colorado. The nonconformable overlying red sandstones and conglomerates are the basal strata of the 25,000-foot-thick Proterozoic Uinta group and were deposited between about 1.4 and 1.0 billion years ago.

PRECAMBRIAN TIME
(545 MILLION TO 4.5 BILLION YEARS AGO)

The beginning of Earth's history is the beginning of Precambrian time. The Archean (Gr. ancient) Eon, from 2.5 to 4.5 billion years ago, and the Proterozoic (Gr. before life) Eon, from 2.5 to 0.54 billion years ago, comprise Precambrian time. With the largest exposure of Precambrian crystalline rocks in the western United States, Colorado is an outstanding place to study the rocks of this time period. The geologic history found in the rocks of Colorado spans nearly 3 billion years; 80 percent of this time falling into Precambrian time.

When discussing this huge amount of time, it is easy to become too casual with the numbers. Remember that the time elapsed between significant intrusive events during the Precambrian, such as between 1.4 and 1.7 billion years ago, is 300 million years—the entire time span of the longest subsequent era, the Paleozoic.

Mountain ranges of vast dimensions appeared repeatedly in Colorado during the Precambrian. It is important to realize that the rock record does not contain the full Precambrian geologic story. Most of the Proterozoic rocks that we observe in Colorado today were formed ten to fifteen miles under the surface. Much of the material that once covered them has eroded away making interpretations of what was going on at the surface more difficult. Nevertheless, observations elsewhere indicate that volcanic activity, mountain building, erosion, and deposition of sedimentary rocks was widespread at the surface during these intrusive and metamorphic episodes.

The Story in the Rocks

Unlike the richly vegetated slopes of today's mountains, Colorado's Precambrian mountains did not have a single plant growing on them. For that matter, neither did the hills and plains because plants had not yet colonized the land. By the end of the Proterozoic, all trace of mountainous topography was gone, flattened into a featureless plain by the forces of erosion. The Precambrian story ends at the Cambrian Period when abundant fossils of shell-bearing animals appeared.

The rocks that form the roots of these ancient Precambrian mountains create some of the most spectacular scenery in Colorado. Although the uplifted mountains of the Front Range are composed of Precambrian rocks that are billions of years old, the present mountains themselves did not form until the Tertiary Period, some 500 million years after the end of Precambrian time.

THE ANCIENT ROCKS: ARCHEAN EON (4.5 BILLION TO 2.5 BILLION YEARS AGO)

The oldest rocks found in Colorado are meteorites that have fallen from other worlds. Many are probably older than Earth itself. Arguably, the oldest, most primitive Earth rocks found at the surface in Colorado are pieces of garnet peridotite from the mantle. Some of this green, garnet-bearing rock formed early in Earth's history as the interior of the planet cooled and segregated into the core and mantle. These rocks were brought from 100 miles deep in the mantle to the surface by the diamond-bearing, kimberlite pipes near the Colorado–Wyoming border.

The oldest crustal rocks in Colorado (the Owiyukuts Complex) are found in a forty-acre parcel in extreme northwestern Colorado along Beaver Creek. Rocks in this area were metamorphosed 2.7 billion years ago. Clearly, the parent rocks were older. The theory of plate tectonics suggests that the metamorphism of these Archean rocks was probably caused by an oceanic plate that pushed beneath an ancient continent of unknown size or shape. The continent must have extended farther into present-day Colorado, but how far? We have no idea.

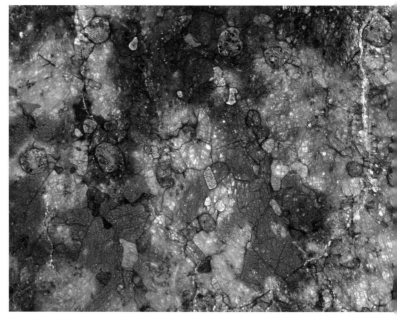

Garnet peridotite from 100 miles deep in the mantle under Colorado

THE ASSEMBLY OF COLORADO: PROTEROZOIC EON (2.5 TO 0.54 BILLION YEARS AGO)

Proterozoic rocks provide evidence that, beginning about 2.3 billion years ago and continuing until 1.0 billion years ago, the embryonic state of Colorado went through an extraordinary sequence of cataclysmic events.

It was just after the beginning of the Proterozoic (about 2.3 billion years ago) that the Archean continent underlying Colorado ripped apart, leaving behind only the small piece of continent in the northwestern corner, plus most of what would become Wyoming. As the part of the continent that had been underlying Colorado moved off to parts unknown, it left in its wake nothing but a deep ocean basin. This new, rifted margin of the continent trended from northwestern Colorado northeastward into Wyoming.

As the newly independent little continent drifted around, elsewhere on the globe the rocks of a volcanic island arc began forming. These independently evolving pieces of crust were probably both being affected by underthrusting, or subduction, of other plates of oceanic crust between about 1.70 and 1.77 billion years ago.

During that time, the island arc continued to evolve with massive granitic intrusion, volcanism, sedimentation, and regional metamorphism. The granitic, intrusive rocks of this age have faint foliation similar to the metamorphic rocks they intruded and are found in most of the mountain ranges of Colorado. The dark gray, granitic rocks exposed west of Boulder are typical of this suite of igneous rocks. A granitic batholith of similar age is exposed on Mount Evans, one of Colorado's fourteeners.

Also during this time, extensive shearing occurred along several northeast trending, linear zones. Some geologists suggest that these zones may have been continent scale faults similar to today's San Andreas Fault. Whether this is true or not, it is well established that the zones severely weakened the crust and were sites of continued faulting during later mountain building episodes. The faults also appear to have strongly influenced the late Mesozoic and Cenozoic mineralization in Colorado a billion and a half years later. Some may still be active.

This 14,255-foot-high peak—Long's Peak, the centerpiece of Rocky Mountain National Park—is part of a 1.42 billion-year-old granitic batholith. The flat top is the remnant of an erosion surface that probably developed less than 23 million years ago.

About 1.7 billion years ago, the island arc (most of future Colorado) was being underthrust from the northwest by the oceanic part of the tectonic plate that carried the Archean continent of Wyoming and northwestern Colorado with it. The underthrusting slowly brought the arc and the small continent closer and closer together until they eventually collided. The suture zone marking this collision between the older Precambrian rocks of northwestern Colorado and Wyoming (2.7 billion years old) and the younger Precambrian rocks of Colorado (1.70+ billion years old) is called the Cheyenne Belt. As a result of this suturing, the continental crust and foundation of modern-day Colorado was finally formed, but that is far from the end of the story.

Proterozoic rocks in the Black Canyon of the Gunnison

For the next 300 million years, not much shows up in the geologic record. Then, about 1.4 billion years ago, widespread granitic magma again intruded the crust throughout Colorado. This intrusive event probably occurred in a magmatic zone parallel to the southeastern coast of the newly merged continent. This granite is familiar to many rock climbers who have inched up the sheer face of "The Diamond" on Long's Peak, the highest peak in Rocky Mountain National Park. As the crust was heated during this intrusive activity, many of the rocks were again metamorphosed and folded.

Most of the Proterozoic rocks were metamorphosed at such high temperatures and pressures that it is not possible to tell what the original rocks were. There are places across the state, however, where the metamorphism was not as intense and the original rocks can be identified. Proterozoic metasedimentary rocks (metamorphic rocks that were originally sedimentary rocks) with features such as graded beds, flame structures, basal scours, and convolute bedding are preserved near Ouray, Gunnison, and in Big Thompson Canyon.

Metaconglomerates are well exposed in Coal Creek Canyon. Metabasalts with vesicular flow tops, pillow lavas, and pillow breccias are found near Vallecito, Gunnison, and Salida. Metavolcanic rocks with textures of ash flows

Outcrop of 1.3 billion-year-old Iron Dike's dark gabbro on Trail Ridge Road. Inset: Map showing location of sixty-eight-mile-long Iron Dike

are also found near Gunnison and Salida. The presence of these types of rocks is significant because all of these original sedimentary and volcanic rocks are typically found in island arcs.

Also about 1.4 billion years ago in the area of the present-day Uinta Mountains of far northwestern Colorado, significant block faulting created an elongate depression that became the site of extensive sediment deposition. For a period estimated to have lasted 450 million years, deposits of red sandstone and shale accumulated. Eventually, these layers formed a deposit five miles thick.

Dikes of dark igneous rocks, such as gabbro, intruded the 1.4 billion-year-old rocks in Colorado when the crust was extended in a northeast–southwest direction. The dikes were probably related to the block faulting occurring in northwestern Colorado. A striking example of these dikes is the sixty-eight-mile-long Iron Dike, which intruded 1.3 billion years ago and extends from just south of Boulder northwestward across Rocky Mountain National Park.

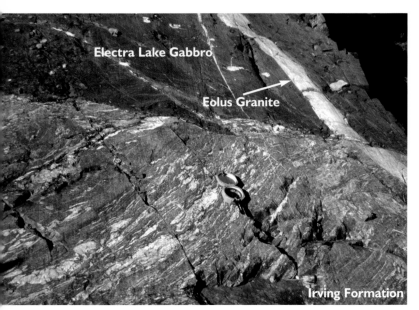

These rocks near Electra Lake north of Durango, document events that occurred during 300 million years of the Proterozoic eon. The Irving Formation originally intruded 1,800 million years ago and was metamorphosed 1,715 million years ago. The dike of Electra Lake Gabbro intruded about 1,400 million years ago and was shortly thereafter cut by the dike of Eolus Granite.

The final intrusive phase of Proterozoic time occurred 1.08 billion years ago when a granitic batholith intruded a 1,300-square-mile area of the Pike's Peak region. The intrusion is composed of several different kinds of granitic rock, but is mostly coarse-grained, pink granite. Igneous rocks of this age in Colorado occur only around Pike's Peak, the Tarryall Mountains, and the Rampart Range.

The end of the Proterozoic is characterized by deep weathering and widespread erosion. It is quite possible that this deep weathering, well displayed in the Pike's Peak Granite near Manitou Springs, is a consequence of the intense global warming that followed the abrupt melting of the worldwide glacial ice about 600 million years ago. During this period, much of the state was eroded to near sea level readying the landscape for the invasion of the Cambrian seas.

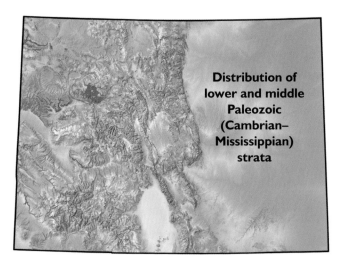

Distribution of lower and middle Paleozoic (Cambrian–Mississippian) strata

Warm, Shallow Seas with Abundant Life: Paleozoic Era (545 to 248 million years ago)

In contrast to the dominance of igneous and metamorphic rocks in the Precambrian rock record, sedimentary rocks characterized the Paleozoic Era in Colorado. Paleozoic sedimentary rocks were deposited in, or adjacent to, relatively shallow seas that invaded the continent periodically through time. Invasion and retreat of the sea and global tectonics weave the geologic story of the Paleozoic Era.

Until geologists had a reasonable explanation for how sea level could vary a thousand feet or more, the cause of the advance and retreat of the seas was hotly debated. Geologists know that ice ages cause significant changes in sea level, but not enough to totally flood the continents. If polar ice sheets were to melt today for instance, sea levels would rise only several hundred feet, enough to flood coastal cities but not the 1,000 feet or more needed in the early Paleozoic to inundate the continent. Although the rock record documents several periods of major continental glaciation during the Paleozoic Era, melting of glaciers doesn't explain the almost complete inundation of the continent by shallow seas that occurred repeatedly during the Paleozoic Era.

The theory of plate tectonics demonstrated that the mechanism for creating major sea level changes is related to the number, shape, and speed of spreading centers in the ocean basins. Spreading centers, like the modern Mid-Atlantic Ridge, are the site of oceanic crustal formation

Note the spheroidal weathering in the red Pike's Peak Granite. The white layers of Cambrian sandstone go up over the spheroidal shapes indicating that they were present when the Cambrian sea inundated this part of Colorado. These spheroidal shapes indicate deep weathering in the late Precambrian.

where molten material moves up and cools and solidifies as tectonic plates move apart. When there are more, and higher, ridges in the ocean basins, there is less room for water. Picture a balloon in the bottom of a large bowl of water. If the balloon is blown up, it displaces the water that spills out of the bowl. Increased volume of the spreading centers displaces seawater out of the ocean basins and sends it flooding across the continents. This indicates that the distribution and character of the sediments being deposited in Colorado during the Paleozoic was strongly influenced by events thousands of miles away on the bottom of the oceans.

Plate tectonics affected Colorado's history in another way. As the North American plate moved around the globe, it carried the state into different climatic zones. Colorado spent the early two-thirds of the Paleozoic in the Southern Hemisphere. By Pennsylvanian time it was at the equator, and by mid-Permian time the northern Horse Latitudes (i.e., 30 to 35 degrees north).

Beautiful exposure of lower Paleozoic strata in Glenwood Canyon

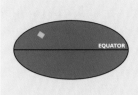

The small globes at the beginning of the discussion of each period show the approximate location of Colorado relative to the equator in a north–south sense. Geologists don't yet have the tools to determine the ancient east–west position of the continents.

The Paleozoic Era is divided into seven periods, each of which is characterized by rocks that reflect the environmental conditions in Colorado at the time of deposition. The early and middle Paleozoic times saw widespread marine conditions, whereas the late Paleozoic witnessed the rise of the Ancestral Rockies and terrestrial conditions. At the end of the Paleozoic Era, North America collided with other continents, and Colorado became part of the supercontinent, Pangaea.

CAMBRIAN PERIOD (545 TO 490 MILLION YEARS AGO)

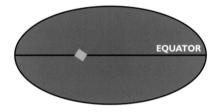

As sea level rose about 500 million years ago, marine waters moved into Colorado from the west and south, depositing sands and pushing the shoreline ahead of the rising sea.

These shoreline and near-shore sandstones are well exposed in the Sawatch Range in Glenwood Canyon, at Baker's Bridge north of Durango, along Highway 24 northwest of Manitou Springs, and near Red Cliff.

During the Cambrian Period, Colorado was close to the equator. The Colorado shoreline separated a land surface with no plants or significant animals from a shallow sea where new organisms evolved rapidly. Organisms that lived in this shallow equatorial sea were preserved as fossils in the Cambrian sandstone. They include clam-like brachiopods, trilobites, and small conical-shelled mollusks.

Precambrian granite

These Cambrian strata were deposited nonconformably on Precambrian granites at Baker's Bridge north of Durango.

In addition to the hard-shelled animals that formed fossils, there were soft-bodied organisms such as algae, jellyfish, and worms, which left little record of their presence. Most of these organisms crawled around on the ocean floor feeding on whatever organic debris they came across. There were not many swimmers yet.

Also during the Cambrian, exotic igneous rocks containing high amounts of thorium and other rare elements were intruded in three complexes in the Wet Mountains between 511 and 535 million years ago. Dikes of an unusual rock, carbonatite, also occur with these intrusions.

During the late Cambrian and early Ordovician periods, igneous dikes were intruded in southwest and south-central Colorado. These dikes occur along a northwest-trending zone, and some are several miles long and as much as ninety-eight feet thick. There are good exposures of these largely mafic (dark-colored) dikes in Black Canyon of the Gunnison near Montrose, Temple Canyon near Cañon City, and in Unaweep Canyon west of Whitewater.

ORDOVICIAN PERIOD (490 TO 443 MILLION YEARS AGO)

EQUATOR

The Ordovician Period began with another invasion of shallow seawater depositing carbonate rocks (predominantly dolomite with some limestone). The sedimentary structures in the lower Ordovician carbonates reveal that this was a long-standing, high-energy, storm-dominated, depositional setting. Features in the rocks indicate that the shorelines oscillated back and forth, occasionally exposing the newly deposited carbonates to air and then covering them again.

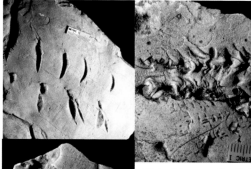

Ordovician trace fossils from the Indian Springs Natural Area—upper left: Trace markings from a sea scorpion; above: Trace markings from a horseshoe crab; left: Trace markings from one of the world's first fish.

The fossils preserved in the Ordovician carbonate rocks are more diverse than those in the Cambrian sandstones; stromatolite-building algae, snails, echinoderms, sponges, crinoids, cephalopods, brachiopods, and trilobites are all common. Ordovician organisms cruised more than the seafloor in search of sustenance. Cephalopods were actually swimmers, and crinoids stretched up so they could capture food several feet off the sea floor. After deposition of the carbonates, the sea withdrew completely for a while but returned much later in the Ordovician.

A period of weathering and erosion preceded a sea level rise in late Ordovician time that deposited sands in central Colorado. Studies of the sandstone in exposures near Florence reveal an astonishing variety of animals living in an estuarine environment. This sandstone contains fossils of one of the earliest known vertebrates—primitive, jawless fish called agnathids.

Also found in this sandstone and younger formations in the Paleozoic strata, are conodonts, which were peculiar, tiny, tooth-like brown fossils. Their origin was a mystery for more than a century until the discovery of imprints of the soft body parts with the tooth-like conodonts in place. This discovery showed that these fossils came from the mouth region of an animal that looked like a modern lamprey eel.

Ordovician carbonate composed of stromatolitic algae near Woodland Park

During the Ordovician Period, North America shifted slowly southward placing Colorado just south of the equator. As sea level rose again just before the beginning of the Silurian Period, a massive amount of gray limestone—composed mainly of the calcareous shells of mollusks, brachiopods, enchinoderms, and corals—was deposited in the warm tropical seas.

SILURIAN PERIOD
(443 TO 417 MILLION YEARS AGO)

Although sea level rose to one of its highest points during the Silurian Period, the extensive sediments that were deposited were soon eroded during a drastic drop in sea level exposing the land. For a long time geologists found no Silurian rocks in Colorado, or even within several hundred miles of Colorado. Then one day someone brought some fossiliferous limestone from northern Colorado to a professor at the University of Colorado and asked him to identify it. The professor realized with a jolt that these rocks came from an area that was thought to have only Precambrian rocks, but the fossils dated to the Silurian Period.

The rocks turned out to be blocks of fossiliferous limestone within diatremes hosted in Precambrian rocks—a very odd juxtaposition. Diatremes form by gaseous explosions in volcanic pipes. These explosions broke off large blocks of overlying rock that fell down into the pipe. Deep

Large fragment of Silurian limestone in kimberlite breccia

underground, the blocks of Silurian rocks within the diatreme were protected from the erosion that removed the Silurian layers still exposed at the surface. These blocks—found only along the Front Range near the Wyoming border—are the only Silurian rock found for 300 miles in any direction. Thanks to the chance discovery and preservation of these unique rocks, there is evidence that the Silurian seas did indeed reach Colorado, and that the volcanic activity that created these diatremes happened during the Devonian Period, 377 to 395 million years ago.

DEVONIAN PERIOD
(417 TO 354 MILLION YEARS AGO)

The withdrawal of the seas during the Silurian Period caused Colorado to be above sea level early in the Devonian Period, and many of the previously deposited rocks eroded. Rocks deposited later in the Devonian Period reflect a relatively slow rise in sea level. Late in the Devonian Period, a marine embayment covered most of the central and southwestern portions of Colorado where first sandstone and then carbonate were deposited. The sandstone is generally well cemented with silica. Because thin shale beds (called partings) made it easy to recognize, geologists named it Parting Sandstone. The sandstone also contains fossil fish and distinctive algae-like beds. The later dolomite contains brachiopods, bryozoans, mollusks, and corals.

Some of the best fossil collecting in Colorado is from the Devonian dolomite beds around the White River Plateau, where the fossils frequently weather out of the rock as almost perfect specimens. These rocks are also associated with significant ore deposits, including lead and zinc, which formed much later in the Tertiary Period.

Overlying steeply dipping, Precambrian strata in Box Canyon Park near Ouray, these Devonian sandstones form a spectacular angular unconformity.

MISSISSIPPIAN PERIOD
(354 TO 323 MILLION YEARS AGO)

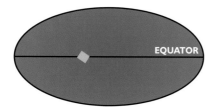

EQUATOR

Sea level peaked for the second time in the Paleozoic Era during early Mississippian time. Colorado was once again completely covered by shallow seas, which resulted in the deposition of massive amounts of gray limestone across the state. About this time (approximately 340 million years ago), North America and Greenland collided with Europe and several other continental pieces to form the northern super-continent Laurasia. Colorado still lay a little south of the equator.

Sea level dropped at the close of the Mississippian Period, and Colorado once again became relatively dry land. During this time, the limestone dissolved and caves formed, creating an irregular land surface known as karst. Weathering of Leadville Limestone formed a soil with a characteristic red color known as terra rosa, literally "red land." This weathered unit lies at the boundary between the strata of the Mississippian and Pennsylvanian periods. Remnants of the ancient soil are best preserved in southwestern Colorado. The Mississippian limestone is a well-known host rock for many Colorado ores emplaced much later during the Cenozoic Era, notably those around the town of Leadville, which lends its name to some of the limestones.

Massive, gray Leadville Limestone overlain by red, weathered limestone breccia called the Molas Formation on Highway 550. The light area in the red material is a remnant of Leadville Limestone that wasn't dissolved, but was left sticking up as a "karst tower."

PENNSYLVANIAN PERIOD
(323 TO 290 MILLION YEARS AGO)

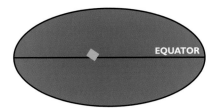

EQUATOR

The story of the Pennsylvanian Period begins with the return of shallow seas depositing marine shales. Although the earliest deposits of this age are fine-grained sands and black shales, conditions changed dramatically after these units were deposited. As the rest of the continent experienced rising sea level, Colorado was dominated by the appearance of two north–south mountain ranges, the Ancestral Front Range and the Ancestral Uncompahgre Mountains.

The mountains rose faster than the rising seas and were attacked by erosion. They shed sediment into basins between the ranges and formed aprons of coarse sediment on both flanks of the two ranges. Since it is rare to find faults and folds formed during this time, the distribution of these coarse sediments is the primary evidence for this particular mountain building event in Colorado.

Examples of the reddish sediment shed off the east side of the Ancestral Rockies are well displayed in the Flatirons near Boulder, Red Rocks, and Roxborough State Parks near Denver, and the Garden of the Gods near Colorado Springs. Rocks shed off the west side of the Ancestral Rockies are prominent between Vail Pass and Avon along I-70. Present-day differential erosion of these rocks created the ledges that characterize the ski runs at Vail. Exposures of the coarse sediment, which record the uplift of the Ancestral Uncompahgre Mountains, appear in the striking red cliffs of Animas Canyon north of Durango, the majestic red cliffs above Gateway, and the gorgeous Maroon Bells near Aspen.

The rising mountains blocked the surrounding seas and isolated the sea between the two ranges into a restricted evaporite basin. Evaporation exceeded water inflow and caused the basin to be saturated by salts, which precipitated as minerals. The stark, desolate appearance of the

Early Pennsylvanian marine shale (lower two-thirds of the slope) in central Colorado was deposited during incursion of marine waters.

hills along I-70 between Eagle and Gypsum is caused by the presence of gypsum and other salts. The evaporites' ability to flow in a solid state caused extensive deformation. The highly contorted bedding in these rocks is easy to see along I-70 near Gypsum. There were also algal mounds growing along the flanks of this evaporite basin, remnants of which can be seen near Minturn and Meeker.

The Paradox Basin in southwestern Colorado is an evaporite basin formed through a similar process. In this basin, thousands of feet of gypsum, salt, and potash were deposited in another restricted arm of the sea. These evaporites are interbedded with limestones and black shales documenting dozens of depositional cycles. Phylloid algal mounds in the area are the reservoirs for some oil fields in the Paradox Basin.

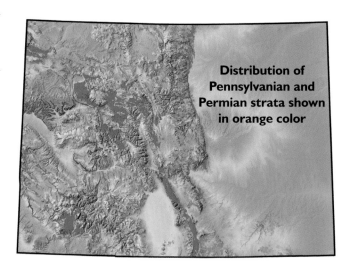

Distribution of Pennsylvanian and Permian strata shown in orange color

Algal mound

Massive Pennsylvanian algal mound typical of those growing around the margins of the restricted sea in central Colorado; note that the beds on the right are all quite thin

Examples of the red sediment shed off the Ancestral Rockies and the evaporite formed in the intervening basin

Pennsylvanian and Permian strata associated with the Ancestral Rocky Mountains—**A:** Evaporite strata in Eagle Valley; **B:** Maroon Formation along I-70 west of Edwards; **C:** Fountain Formation near Lyons; **D:** Maroon Formation in Maroon Bells (Snowmass Mountain in the distance); **E:** Cutler Formation along the Animas River north of Durango; **F:** Weber sandstone at Echo Park in Dinosaur National Monument

Although the Ancestral Rockies uplifted much of the area, shallow seas still repeatedly invaded the lowland areas. Marine fossils document as many as twenty marine cycles in the strata that outcrop near Vail and also in the strata in the subsurface in the Denver Basin in eastern Colorado. These cyclic deposits are the result of continental glaciers that repeatedly stored and released large volumes of water causing the relatively rapid fluctuation of sea level during the Pennsylvanian Period. These glaciers formed on Gondwana, the continental mass that included Africa, South America, Australia, and other continental pieces situated astride the South Pole. Oddly enough, this means that glaciers in Australia directly affected the type of rock that was deposited in Colorado.

PERMIAN PERIOD
(290 TO 251 MILLION YEARS AGO)

As the Permian Period began, erosion continued to wear away the Ancestral Rockies and fill the intermountain basins with sediment. The humid conditions of Pennsylvanian time were replaced by a more arid climate. As sea level dropped, the retreating shallow sea exposed more and more dry land in eastern Colorado. On the land, large dune fields dominated the newly exposed landscape. In the western part of the state, limestone and shale were still being deposited in a shallow sea.

As the Permian Period ended, the Ancestral Rockies were reduced by erosion and Colorado was a relatively flat, low-lying region with an arid or semi-arid climate. All of the continents had coalesced to form the supercontinent of Pangaea, and Colorado was on the edge of the shallow sea with sea level continuing to drop. Few fossils of the animals and plants that inhabited Colorado during the Permian are found, although tracks of early reptiles have been found in sandstones.

Ancient sand dunes; forest beds of Permian dune sandstones in the flagstone quarry above Lyons

THE END OF AN ERA: MASS EXTINCTION

The Paleozoic Era began with an explosion of life and ended with a profound mass extinction. The extraordinary increase of species diversity and complexity that occurred during the first 10 to 20 million years of the Paleozoic Era is so remarkable that it is referred to as the Cambrian Explosion. In stark contrast to its vibrant beginning, at the close of the Paleozoic at least half of the known families of both marine and terrestrial organisms died out. Scientists estimate that in this period of mass extinction, 75 percent of the amphibian families and more than 80 percent of the reptile families disappeared. Few paleontological questions have attracted more attention than the search for the cause of this disaster.

Land of Dinosaurs: Mesozoic Era (251 to 65 million years ago)

The Mesozoic Era embraces three distinctly different periods: Triassic, Jurassic, and Cretaceous. The shape and location of continents, and the nature of life forms on Earth, changed greatly during Mesozoic time. During the first two periods, Colorado was a land of low relief with a persistent warm, dry climate. During the third and final period, a vast sea covered the entire state, but retreated as mountains began to rise at the beginning of the Laramide mountain building event.

The Mesozoic Era is popularly known as the Age of the Dinosaurs, and Colorado was home to many of them. During this period, dinosaurs evolved from small reptiles into behemoths that ruled land, sea, and air. This domination ended dramatically with their sudden and complete extinction at the end of the Mesozoic Era. Although small primitive mammals and other vertebrates made their first appearance in the fossil record in early Mesozoic time, they lived in the shadows of the giant reptiles.

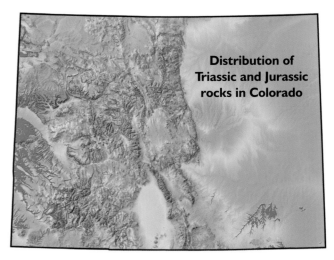

Distribution of Triassic and Jurassic rocks in Colorado

Mesozoic rocks lie at, or very near, the surface of about 35 percent of Colorado and under an even greater area in the subsurface. These rocks contain important energy resources including oil, natural gas, and coal. Several of the formations also contain important industrial minerals and groundwater as well as uranium, radium, and vanadium in western Colorado.

Exposure in central Colorado—brick-red shales and siltstones of Triassic age; purple and gray shales of the Jurassic Morrison Formation; and rimrock of Cretaceous Dakota sandstone

Triassic Period
(251 to 206 million years ago)

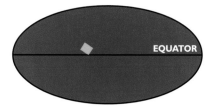

EQUATOR

A superb place in western Colorado for viewing Permian and Triassic red beds is along State Highway 141 south of Gateway.

Triassic red beds impart hues of red, maroon, and pink to many western Colorado landscapes. They are fine-textured sandstones, siltstones, and shales that were deposited in mudflats, on alluvial plains, and in dune fields adjacent to eroding highlands. The sediments deposited during much of the Triassic Period are remarkably similar in their composition and general character throughout the United States.

The red beds are sometimes interfingered with limestone, halite, and gypsum—probably the result of brief excursions of restricted marine waters from the north onto flat-lying coastal and alluvial plains. Triassic red beds are exposed along much of the eastern edge of the Front Range uplift and form part of a "strike valley" between the two steeply tilted, hogback ridges. These valleys extend intermittently from Douglas County north to the Wyoming state line. In western Colorado, the Triassic red beds are more easily eroded than most other formations and generally appear as debris-covered slopes between resistant ledges.

Although fossils are rare in Triassic rocks, tracks of small bipedal dinosaurs, such as *Coelophysis*, have been found at several localities in Colorado. Bones of this dinosaur were discovered near New Castle in 1998. In addition, partial body fossils have been found in western Colorado for two different types of reptiles with thecodonts (vertebrate teeth set in sockets). These finds include teeth from a phytosaur, a primitive twelve-foot-long crocodile-like predatory reptile that inhabited lakes in late Triassic time, and a few armor plates from a land-roving herbivore reptile called an aetosaur. Three locations in western Colorado have yielded trackways of different Triassic animals.

Big Horn sheep hang out on Triassic rocks, which are a good source of natural salts in Dinosaur National Monument.

An excellent place to observe the change from Triassic red beds to Jurassic cross-bedded sandstones is in Colorado National Monument near Grand Junction.

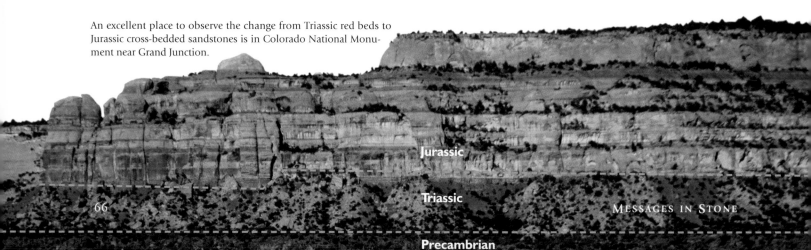

Jurassic

Triassic

Precambrian

Jurassic Period
(206 to 144 million years ago)

Deposition of red beds in Colorado was followed in late Triassic and early Jurassic time by deposition of widespread dune sands in a warm, desert climate. This climate change resulted from the separation of the North American plate from Africa. Hooked to North America, Colorado traveled north, out of tropical latitudes into Horse Latitudes. The separation also changed land and sea configurations of the newly separated continent and its adjacent seas.

The Jurassic sandstones with large-scale cross bedding are preserved sand dunes. These ancient dune sandstones form spectacular light-colored cliffs throughout much of extreme western Colorado. Deposition of dunes was interrupted four times during the first 45 million years of the Jurassic Period when marine waters advanced from the

Typical rounded pink and white appearance of the early Jurassic Entrada dune sandstone, along Highway 550 south of Ridgeway

Colorado Geological Survey scientist points out dinosaur tracks on CGS' annual mountain-bike trek called "Dash for the Dinosaurs"

northwest. Some of these marine rocks contain fossils such as ammonites, belemnites, and pelecypods (clams). This seaway and its deposits covered parts of northwest and north-central Colorado. Remnants of the Ancestral Front Range and Uncompahgre highlands remained just above sea level in the south and central areas of the state.

After final retreat of the Jurassic sea, the entire Rocky Mountain region became a vast continental lowland of lakes, swamps, dunes, and braided and meandering streams. The resulting deposits are known as the Morrison Formation, which is amazingly similar in appearance and thickness throughout this region. This great lowland area and its deposits covered a ten-state area in the western interior of the United States and southern Canada.

The Morrison Formation yields one of the richest fossil assemblages of dinosaurs on the continent including *Stegosaurus* (the state fossil of Colorado), *Diplodocus*, *Apatosaurus* (formerly known as Brontosaurus), *Camptosaurus*, *Camarasaurus*, *Ceratosaurus*, and *Allosaurus*. The Colorado Morrison Formation also provides the continent's largest assemblage of dinosaur trackways. The trackway area lies south of La Junta, along the Purgatoire River, and has more than 1,300 tracks and 100 trackways in exposed limestone layers of the Morrison Formation.

CRETACEOUS PERIOD
(144 TO 65 MILLION YEARS AGO)

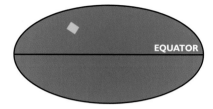

The Cretaceous Period began in Colorado with the lowland topography of the Jurassic while the Gulf of Mexico approached from the south. Sea level rose and the Arctic Ocean migrated south into Colorado. Widely fluctuating geologic conditions resulted in a complex intertonguing of marine and terrestrial deposits during this period. For much of the Cretaceous, marine waters continually attacked Colorado's landscape, alternately advancing and retreating from different directions.

Eventually, seas advancing from the north and south joined and covered most of Colorado with the Western Interior Seaway. Uplift and volcanic activity to the west in Utah contributed sediment pulses and ash fall deposits. Many minor fluctuations of sea level also produced pulses of marine deposits, and the entire area from central Utah through western Colorado was a slowly subsiding foreland basin allowing thick accumulation of sediment. The

Thick Cretaceous sandstones form steep cliffs. The La Plata Mountains in the background are composed of younger Cretaceous shales intruded by Tertiary igneous sills, Mesa Verde National Monument

basin subsidence also affected the shifting of the shorelines, further complicating the battle for dominance between terrestrial and marine deposition. The beginning of the Laramide mountain building event ushered in the final retreat of marine waters and a return to terrestrial conditions that began during the last few million years of the Cretaceous Period and continue today.

Cretaceous rocks represent four basic depositional environments. Sediment deposited in deep- to moderate-depth marine waters yielded limestone and chalk. Sediment deposited into shallow to intermediate depths of marine waters produced dark shale with interbedded thin limestone, siltstone, and sandstone. Massive sandstone formed near shorelines or in intertidal areas. Sediments in coastal plain, lagoon, swamp, and alluvial plain environments became claystone, shale, lenticular sandstone, coal, and conglomerate.

At this time, Colorado was subtropical. Bees, wasps, and butterflies appeared and flowering plants developed. Palm trees were common. Cretaceous rocks of eastern and western Colorado are similar in many respects, but quite different in detail.

Cretaceous shales form extensive badlands in western Colorado

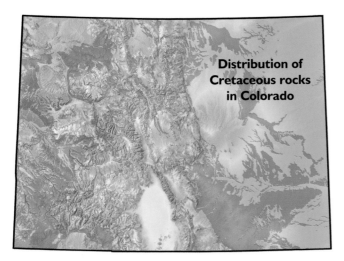

Distribution of Cretaceous rocks in Colorado

The oldest Cretaceous rocks in Colorado are the economically important Dakota Group that were widely deposited as a complex of beach, deltaic, estuarine, and minor near-shore marine sediments. Rocks from this group are found throughout Colorado.

Sandstones from the Dakota Group are important oil, gas, and groundwater reservoirs. These sandstones are resistant to erosion and form the backbone of the Dakota Hogback ridge along the eastern foothills of the Colorado Front Range. The hogbacks are steeply dipping sedimentary rocks that were upturned during the Laramide mountain building event. Where Dakota sandstones are relatively horizontal they typically form a rimrock or a distinctive mid-slope ledge.

As the seaway deepened and marine waters advanced over the deposits of the Dakota shoreline, fine-textured limestone and shale were deposited in eastern Colorado. Localized deepening of the seaway followed, and shallow marine sediments were succeeded by marine mud deposited in deeper waters farther from shore. The mud, thousands of feet thick, became the shale of the Pierre Shale Formation in eastern Colorado. Pierre Shale is present under all of eastern Colorado and is as much as a mile-and-a-half thick near the mountain front. The shale forms low relief areas because it is easily weathered and eroded. In western Colorado, thick deposits of mud became the widespread Mowry and Mancos shales.

At the time that the Pierre Shale was deposited in eastern Colorado, the interbedded, terrestrial, and marine strata of the Mesaverde Group and Mancos Shale were deposited in the western part of the state. Extensive deposits of coal, resulting from the accumulation of organic matter in marshes and lagoons, are widespread in the Mesaverde Group.

Beginning about 70 million years ago, the long history of deposition and subsidence of the Mesozoic Era came to an end with one of the significant events in Colorado's geologic history—the Laramide mountain building. At this time the entire western continental interior rose regionally and the crust was broken into blocks of all sizes. Some of the largest blocks rose to become mountains while others dropped to form basins. The seas of the late Cretaceous Seaway retreated for the last time and were replaced by the emerging mountain ranges and terrestrial basins that would dominate the ensuing Cenozoic Era. Along the margin of the retreating sea, sand beach and bar deposits in eastern Colorado formed the Fox Hills sandstone and the overlying, interbedded sand and coal of the Laramie Formation. This pattern of sandstone and shale interbedded with coals characterizes Upper Cretaceous rocks in the basins throughout Colorado.

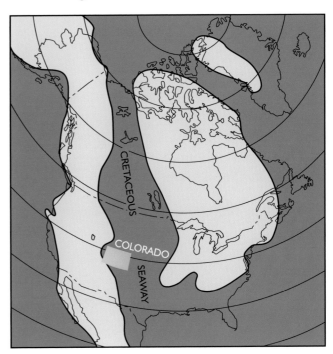

Seas repeatedly inundated Colorado from the south and north causing shorelines to shift back and forth across the state

Hurricanes in Colorado

Only people in coastal areas worry about hurricanes, so hurricanes may seem a peculiar topic in a book about Colorado geology, but, during the Cretaceous Period, Colorado did have coastal areas, although there were no people, of course. Studies of the Cretaceous sandstones of the Mesa Verde Formation in northwestern Colorado reveal beach and tidal flat deposits. The preserved strata have hurricane washover channels and fans identical to those formed during hurricanes along the Gulf Coast today.

The hurricanes of today usually begin in the Atlantic and build in intensity as they enter the Gulf of Mexico heading for land along the Gulf Coast. Imagine living 100 million years ago when hurricanes could build strength through 1,300 additional miles of ocean in the Cretaceous Interior Seaway before reaching the beaches of northwestern Colorado. It would have been bad enough having all those monster dinosaurs running around, but then to have monster hurricanes on top of that.

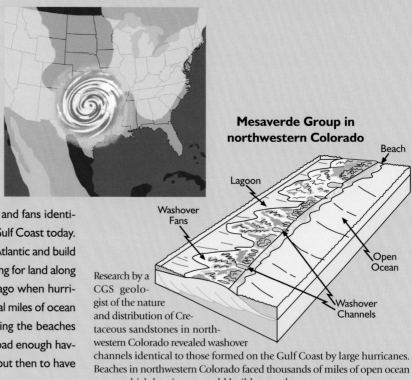

Mesaverde Group in northwestern Colorado

Research by a CGS geologist of the nature and distribution of Cretaceous sandstones in northwestern Colorado revealed washover channels identical to those formed on the Gulf Coast by large hurricanes. Beaches in northwestern Colorado faced thousands of miles of open ocean across which hurricanes could build strength.

THE GREAT WIPE-OUT: K/T BOUNDARY

In the 1940s, South Table Mountain west of Denver was the first place in the world where the boundary between the Cretaceous and Tertiary Periods was described in terrestrial rocks with only dinosaur bones below and only mammals above. This site is so significant that the National Science Foundation held its fiftieth birthday party on the outcrop at South Table Mountain.

After decades of speculation, a hypothesis was proposed suggesting that a meteorite with a diameter slightly smaller than the width of Denver traveling at 50,000 to 100,000 mph, slammed into Earth causing extinction not only of the dinosaurs but many other animals and plants as well. The impact on Mexico's Yucatán Peninsula left a crater more than ninety miles in diameter. The ejecta (pulverized and melted rock resulting from the impact) were launched high into Earth's atmosphere, circled the globe, and blocked sunlight, turning day into night.

As the ejecta rained down, friction heated the debris. The atmosphere turned into an oven, baking the ground and everything on it, and igniting the vegetation and forests. Soot from the fires added to the ejecta cloud that shrouded Earth. The planet's surface was changed in an instant. The Age of Dinosaurs (Mesozoic) became the Age of Mammals (Cenozoic).

The ejecta cloud, composed of shocked mineral grains, micro-tektites, and rock fragments, blanketed Earth and settled to the surface forming a thin layer. This blanket was preserved only where conditions were perfect, and conditions in southern Colorado were ideal. Colorado's Cretaceous shoreline along the Cretaceous Interior Seaway looked similar to present-day Louisiana, with deltas, mudflats, and swamps. The K/T boundary clay was preserved in the calm, non-agitated waters of the swamps that later became the rich coal deposits of the Raton Basin in southern Colorado.

The thin K/T boundary, containing the fall-out from the asteroid impact, is between the white clay layer and the black coal, north Clear Creek locality south of Trinidad

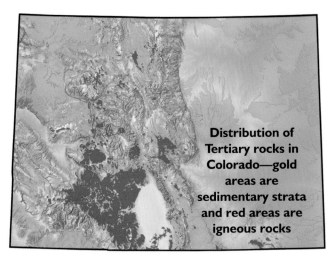

Distribution of Tertiary rocks in Colorado—gold areas are sedimentary strata and red areas are igneous rocks

Colorado rocks and geologists played key roles in proving the impact hypothesis, finding the crater, and proving that a crater in Yucatán was the source of the impact debris. Southern Colorado's dozen sites that preserve the K/T boundary layer are so important that the Smithsonian Institution collected a two-and-a-half-ton sample of the K/T boundary from south of Trinidad and has it on display in Washington, D.C.

The layer yields anomalously high amounts of iridium, a rare element in terrestrial rocks, but common in meteorites. The existence of the iridium anomaly at the K/T boundary in coal-swamp deposits in Colorado proved that it was not just some phenomenon created by seawater. Indeed, the highest iridium content ever measured in continental rocks was in Colorado rocks south of Trinidad. Below the layer is a great diversity of fossil organisms; above the layer is a marked decrease in abundance and diversity of fossil organisms of all sorts, not just dinosaurs.

The layer is commonly composed of clay particles, with or without soot, and fragments of minerals that show evidence of being strongly shocked by impact. The size of the shocked fragments and the existence of two layers in Colorado sites also indicated to scientists that the impact must have been located near North America. Most of the chemical data that finally convinced scientists that the Chicxculub crater on the Yucatán peninsula was the source for the K/T deposit also came from Colorado rocks.

A good place to observe the K/T layer is in Trinidad Lake State Park where an informational sign sits along side the Long Canyon trail.

Age of Mammals: Cenozoic Era (65 million years ago to present)

In comparison to those eras that preceded it, the Cenozoic is relatively short and is divided into two very unequal periods of time—the Tertiary and the Quaternary. Sixty-three million years are classified as the Tertiary Period, while only the last 2 million years belong to the Quaternary Period. Since the Cenozoic Era is the most recent span of geologic time, much is known about it. Mammals, such as rhinoceros, mammoths, and camels, eventually replaced the dinosaurs of the Mesozoic Era. The animals and plants that evolved during the Cenozoic are the ancestors of the fauna and flora of our present environment.

Mountain building, rifting, volcanoes, glaciers, widespread erosion, and basins filled with sediment characterized the era in Colorado. The Rockies are not Precambrian mountains, but rather Cenozoic mountains made of Precambrian rocks.

With the death of the dinosaurs, mammals flourished on Earth. Shown here is a juvenile mammoth.

Mountains, Erosion, and Volcanoes: Tertiary Period (65 to 1.8 million years ago)

The Tertiary Period in Colorado includes two mountain building events, three igneous events, and deposition of thick terrestrial sediments. The Laramide mountain building event, which began in the late Cretaceous, continued for the first 25 million years of the Cenozoic Era. The tens of thousands of feet of uplift and downwarp formed the ranges and basins of much of Colorado's spectacular, present-day scenery. Dramatic products of this period include the fold and fault belt from Lyons to Fort Collins, Boulder's Flatirons, the hogbacks along the Front Range, Garden of the Gods in Colorado Springs, the Hogback Monocline near Durango, and the Grand Hogback of Glenwood Springs. The monoclines of Colorado National Monument in Grand Junction and Dinosaur National Monument of northwestern Colorado also emerged at this time.

Dinosaur National Monument is a special place to observe and study huge Laramide folds and faults. The Mitten Park structure starts as a fault at the level of the Green River. Higher up, the fault dies out into a continuous fold. The vertical scale of the fold is approximately one-half mile. Numerous monoclines are easily accessible here.

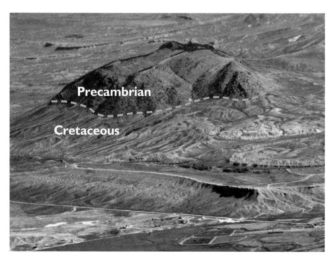

Aerial view of a Laramide thrust fault at Wolford Mountain near Kremmling that places Precambrian granite over Cretaceous rocks, the Colorado River is in the foreground. This thrust fault is about sixty miles long and forms the western boundary of the Front Range. It may have had as much as five miles of horizontal displacement in places.

Close-up of the thrust contact near Keystone. Here Precambrian granite (light) rests on Cretaceous shale (dark) baked by a Cenozoic intrusion. In looking at this knife-sharp contact, it is amazing to consider that the granite may have moved five miles before it came to rest on the shale.

Right: Sixty-four-million-year-old basalt-like flows of shoshonite cap the top of North and South Table Mountains, looming above the Coors brewery in Golden. Like most of the volcanoes of this age in Colorado, nothing remains of the volcanic cone except its intrusive roots near Ralston Reservoir.

Below: Aerial view looking south showing the Ralston Dike (red), which is the root of the volcano out of which the lavas in the Table Mountains flowed 64 million years ago. The orange bands and the white tops of the mountains indicate the distribution of the volcanic flows. Note that there are at least three flows in North Table, but only two flows in South Table, supporting the idea that they came from the dike to the northwest.

South Table Mountain

North Table Mountain

Flow 3

Flow 2

Flow 1

Ralston Dike

This imposing landmark, Pulpit Rock in Colorado Springs, holds key information about the growth of the Laramide-age mountains. In the lower slopes are sandstones with an abundance of volcanic fragments, which tells geologists that many volcanoes were around even though we can't find them today. As you go up the slope you find more and more granite fragments that indicate erosion in the mountains had removed all of the volcanic rocks and was cutting down into the Pike's Peak Granite. The volume of sedimentary material, its coarseness, and its composition record the timing and nature of the Laramide mountain building event.

Middle Cenozoic sedimentary rocks on the high plains. The Laramide mountains were worn away by erosion, transported, and dumped as debris out onto the plains.

Middle Cenozoic basin-fill deposits of Middle Park west of Granby consist of a mixture of sedimentary and volcanic rocks.

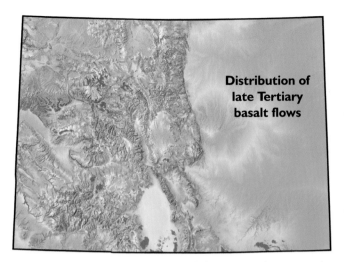

Distribution of late Tertiary basalt flows

The San Luis Valley is a prominent expression of the Rio Grande Rift. The Sangre de Cristo Range bounds the valley on the east. Vertical movement on the fault zone between the valley and range during the late Cenozoic extension is calculated to be approximately four miles.

Igneous activity accompanied the Laramide mountain building event. Intrusive rocks of Laramide age (72 to 50 million years old) lie throughout the northeast-trending Colorado Mineral Belt. Although volcanoes are not preserved, we know that magma also reached the surface and built volcanoes because we find volcanic fragments in the sedimentary rocks of the same age in the Denver Basin and we have the preserved flows on North and South Table Mountains.

Laramide mountain building ended much as it began, with activity tapering off at different times in different places. About 38 million years ago, a combination of beveling by erosion and burying by deposition reduced the entire area to a broad undulating surface of low relief. Former mountains were buried in their own debris. The period of erosion during and following the Laramide is recorded in thick sequences of Tertiary sediments in the basins between mountain ranges and on the plains. Although the post-Laramide erosion surface has been uplifted, faulted, and dissected by streams in ensuing years, remnants of the eroded surface are preserved in many areas of south-central Colorado.

After a quiet period of nearly 2 million years, Colorado once again erupted and became a land of fire. Widespread igneous activity broke out about 36 million years ago and continued for 10 million years. Volcanic rocks were deposited in many areas of the state on the low-relief, post-Laramide erosion surface. Volcanic material

originating as hot, incandescent ash flows and lahars (volcanic mudflows) flowed through the broad, well-developed stream channels of the low-relief erosion surface to distribute deposits far beyond the volcanic source centers. At its maximum extent, this great coalescing sheet of volcanic outpourings covered half of the state.

The last major episode of tectonic and volcanic activity began in the latter half of the Tertiary Period. Unlike the shortening of the preceding Laramide mountain building event, this uplift is characterized by spreading apart in the southern Rocky Mountains. The previous andesitic igneous activity was replaced by basaltic volcanism as Colorado's crust pulled apart and extended to the east and west. As extension progressed, the crust was broken into many blocks and basaltic magma from Earth's mantle rose up through the faults and poured out on the surface. The extension also caused blocks of crust to rise and fall creating mountains and basins. This activity continues and is evidenced by active faulting with accompanying earthquakes and fairly recent basaltic eruptions.

Much of present day topography is a result of the block-faulted mountains and basins. During the late-Tertiary block faulting in Colorado, some new faults were formed, but many were reactivated older Precambrian, late Paleozoic, or Laramide structures. Individual fault displacements of at least 20,000 feet occurred between the San Luis Valley and the Sangre de Cristo Range during the Late Cenozoic.

The most dominant late-Tertiary structural feature in Colorado is the Rio Grande Rift and the accompanying uplifted ranges that bound the down-dropped valleys. The Rio Grande Rift can be traced nearly 500 miles from Mexico through west Texas and New Mexico, entering Colorado through the north-trending San Luis Valley, and continuing northward in a series of segments to the vicinity of Steamboat Springs. The rift is a series of north-trending down-faulted basins or grabens, flanked on one or both sides by up-faulted mountain ranges. The most outstanding grabens or structural valleys of the rift are the San Luis Valley, Upper Arkansas Valley from Salida to Leadville, and the Blue River Valley from Silverthorne to Kremmling. Examples of adjacent faulted mountain blocks include the Sangre de Cristo, Mosquito, and Williams Fork Ranges on the east flank of the rift and the Sawatch, Tenmile, and Gore Ranges on the west. The subsiding grabens acted as sedimentary basins and collected thousands of feet of debris that was shed as the rising mountains were increasingly subjected to the forces of erosion.

At the end of the Tertiary Period, continued uplift accelerated canyon cutting and erosion. Royal Gorge, Clear Creek, Glenwood, Big Thompson, Cache la Poudre, and the Black Canyon of the Gunnison are a few of the spectacular canyons cut during the last 5 million years. The Colorado River formed Glenwood Canyon as it eroded into the southern flank of the White River Uplift. Most of the steep inner walls of the canyon were carved during the past 3 million years. At the same time, the Colorado River was carving Glenwood's larger sibling, the Grand Canyon, downstream in Arizona. Hand-in-hand with stream erosion, uplift played a role in canyon formation, and probably continues today.

This 1,000-foot gorge, Royal Gorge west of Cañon City, in Precambrian rocks is spanned by the world's highest suspension bridge. The Arkansas River carved the narrow gorge in about 3 million years. From here, the river soon enters the plains and wanders more than 1,000 miles to join the Mississippi River.

AND NOW, THE QUATERNARY PERIOD (1.8 MILLION YEARS AGO TO PRESENT)

The geologic period in which we live started 1.8 million years ago. One of the abiding mysteries about the Quaternary is why many large mammals—giant bison, mammoths, camels, ground sloths, and saber-toothed cats—became extinct approximately 10,000 years ago.

Paleontologists have considered many theories to explain why this time was so deadly. It is possible that the extinctions hinged on the climate. We know that dramatic changes occurred around this time. Scientists speculate that this change may have disrupted breeding cycles and/or left animals more vulnerable to disease. Perhaps Paleo-Indian hunters, locked in their own struggle for survival, killed so many animals that the species couldn't survive.

In all probability, the causes for the extinction lie in a combination of factors. We may never know exactly what happened. What we do know from the rocks of the period

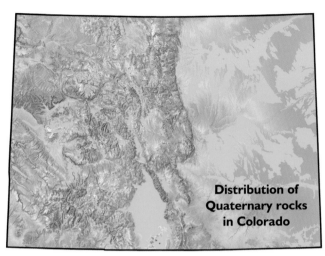

Distribution of Quaternary rocks in Colorado

is that the Quaternary was a time of monster glaciers, moving bodies of ice thousands of feet thick that left indelible marks on the landscape. The power of glaciers, water, wind, gravity, volcanoes, and earthquakes etched the surface into the awe-inspiring landforms we see today.

The moraines were deposited during the Little Ice Age, probably in the past two hundred years in the Indian Peaks of the Front Range. This is the southernmost glacier in the Rocky Mountains, Arapahoe Glacier.

Precambrian Metamorphic Rocks

Glacial Ice

Snow

Snow

End Moraine

Land of Ice: Quaternary Glacial Geology of Colorado

A glacier is a moving body of ice formed by accumulation, compaction, and recrystallization of snow. Before their disappearance 12,000 years ago, large glaciers thousands of feet thick filled valleys and left their marks on the mountainous landscapes of Colorado. Although the massive glaciers are long gone, their shadows linger on the land in the valleys they reshaped, the lakes they created, and the stones they polished in their icy progress. More than a dozen modern glaciers are indicated on topographic maps in Colorado. Today's small glaciers are not remnants of the large glaciers, but were fortunate enough to form in sheltered mountainous landscapes in Colorado during the Little Ice Age between 1200 and 1880 A.D.

White snow caps the gray ice mass, Andrews Glacier, in Rocky Mountain National Park

Geologists are able to distinguish deposits in Colorado of at least three ice ages during the Quaternary Period. The two most recent glacial events peaked between 130,000 and 150,000 years ago and about 20,000 years ago. The age of older Quaternary glacial deposits is harder to decipher in the state.

This huge crack (crevasse) in the ice mass of Andrews Glacier demonstrates that it is moving.

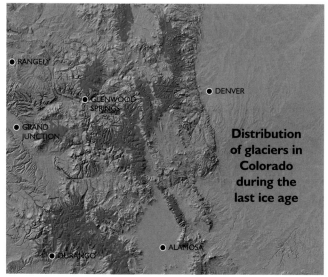

Distribution of glaciers in Colorado during the last ice age

Glacial Sculpting

Since glaciers flow, they erode, transport, and deposit rocks and sediment as they move down valley. When water penetrates cracks and joints in the rock over which glaciers flow, the meltwater freezes and expands, breaking loose blocks of rock and sediment. The rock and sediment frozen in the base and sides of glaciers act like the sand in sandpaper—grinding, polishing, scratching, and rounding the rocks under the glacier. In the process, the transported rocks also become smoothed and rounded.

A good example of glacial sculpting is a roche moutonnée, which is an asymmetrical, elongate knob or hillock of resistant bedrock that has been smoothed and scoured by moving ice on the up-glacier side. On the down (lee) side, the rock is steep and hackly from glacial quarrying.

Polished, grooved, and striated bedrock near Isabelle Glacier in the Indian Peaks Wilderness

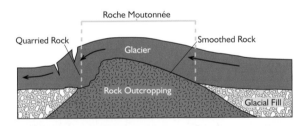

A roche moutonnée in the middle of Moraine Park in Rocky Mountain National Park, with the south lateral moraine forming the backdrop

Polished and striated boulder in the Arapahoe Moraine

Rounded and sculpted Precambrian rock near Isabelle Glacier, Indian Peaks Wilderness

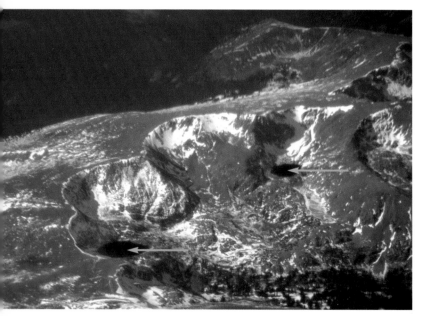

In a situation common in Colorado's glaciated regions, a series of cirques all feed into one valley, in this case, near Berthoud Pass. Note the tarns indicated by arrows.

Glacial U-Shaped Valleys and Cirques

Few glacial landforms have caught the imagination of geologists and non-geologists alike more than the glacial cirque. A cirque is a bowl-shaped, amphitheater-like basin formed by a large glacier at the head of a valley. The small glaciers that exist in the lofty mountains of Colorado occupy cirques excavated thousands of years ago by their large predecessors. Small lakes, called tarns, occupy many of the glacially scoured basins.

Stream-cut alpine valleys are characteristically V-shaped. As a glacier moves out of its cirque down a valley once occupied by a stream, the glacier widens, steepens, and straightens the valley. The valley is transformed into a glacial, U-shape. Good examples of U-shaped valleys and cirques are abundant in the Colorado Rockies.

The unusual relationship between topography and the dipping sedimentary strata accentuates the amphitheater shape of Horseshoe Cirque, visible from Fairplay, in the Mosquito Range.

Valleys that were glaciated commonly have a U-shape. This U-shaped valley in the San Juan Mountains holds Emerald Lake, one of Colorado's largest natural lakes.

The Little Matterhorn (11,586 feet) in Rocky Mountain Park was formed from glaciers eroding away at three sides.

Glacial Arêtes and Horns

Where two glacial cirques cut back toward each other, a jagged, knife-sharp ridge of bedrock called an arête may be all that is left between them. Peaks that are, or have been, surrounded by glaciers tend to have a characteristic form and steep straight slopes. Where isolated, such mountains may form horns that have three or four distinct faces. Sometimes arêtes link horns.

The backward erosion of the walls of adjoining cirques often leave a knife-edge, jagged ridge (arête) such as this one in the Tenmile Range.

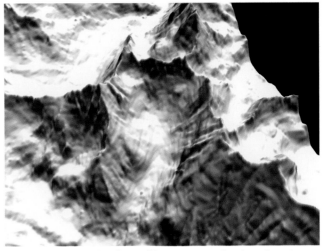

Digital elevation model of part of the Gore Range showing arêtes, cirques, horns, and U-shaped valleys. How many can you find?

Answers: 8 cirques, 8 arêtes, 5 horns, and 2 U-shaped valleys

Glacial Deposits

A glacier carries all sizes of rock materials at its base, sides, and surface, and it deposits the material along the sides and floor of the valley in which it moves. Till is the general term for the ill-sorted mixture of fine and coarse rock debris deposited directly from the glacier ice. Erratics are isolated boulders left on a glaciated surface.

Glacial till above Turquoise Lake, west of Leadville

This deposit is in one of the end moraines in the city of Durango. The glacier that deposited the material started forty miles away in the San Juan Mountains.

Rocks carried by the glacier were deposited when the ice melted away, leaving an assortment of boulders (erratics) that are different from the bedrock upon which they sit, Rocky Mountain National Park.

Glacial Moraines

The most obvious landforms composed of till are moraines. Moraines can be undulating mounds or sharp ridges depending on how long a glacier remained stable in a particular position or how much erosion has taken place in the time between deposition and the present.

Lateral moraines form on the sides of a glacier and merge with an end, or terminal moraine. The end moraine forms when ice stabilizes for a time prior to retreat. An end moraine may dam melt-water to create a lake on its up-valley side. Turquoise Lake near Leadville is dammed by an outstanding example of an end moraine. A series of recessional moraines—up-valley from the end moraine—may be laid down as the glacier recedes and re-stabilizes in its final retreat. The Animas Glacier, one of the longest in Colorado, formed well-developed end moraines in Durango.

A digital elevation model and satellite image of Turquoise Lake west of Leadville, dammed by a textbook example of an end moraine.

Moraine Park in Rocky Mountain National Park is a wonderful place to observe glacial landforms. The south lateral moraine is more than 600 feet tall. A roche moutonnée is indicated by the red arrow. The Twin Sisters are the high peaks on the left, and Estes Cone is the peak on the right.

The Story in the Rocks

Anatomy of a Glacial Valley: Fall River Valley, Rocky Mountain National Park

A

Cirque where the ice accumulated and then flowed down the valley nine miles to Horseshoe Park, which is 3,800 feet lower in elevation. Arrow is Alpine Visitor Center

B

Polish and striations on granitic rocks along Fall River Road, which winds its way up to the Alpine Visitor Center.

C

Rounding and horizontal grooving of valley walls along Fall River road

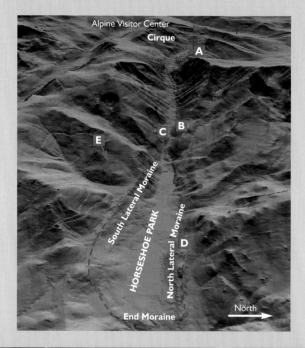

Alpine Visitor Center
Cirque
A
B
C
E
South Lateral Moraine
HORSESHOE PARK
North Lateral Moraine
D
End Moraine
North

E
HP
SLM
SLM

View of south lateral moraine (SLM) and Horseshoe Park (HP) from Rainbow Curve. Small red dot is in a bog created by the lateral moraine blocking drainage from the south. Fall River meanders through Horseshoe Park. The park is flat bottomed because the end moraine dammed the valley and sediment filled the depression during glacial retreat.

D

View northward across Horseshoe Park of lateral moraine on north side of the valley

Periglacial Quaternary Landforms

Periglacial features are associated with glacial landforms, although in many cases they are located far from glaciers. The term "periglacial" is applied to the processes and landforms (regardless of age) associated with very cold climates in areas not permanently covered by snow or ice. In Colorado, the features are common at high elevations (above 11,500 feet) where prevailing temperatures are so low that the ground remains frozen for much of the year. The effects of repeated freezing and thawing, and the growth of ice masses in the ground, produce many unique alpine features.

An interesting effect of ice-crystal growth in soils (ground ice) is the movement of soil and rock fragments toward the surface where they form mounds and rows of soil or rock. Where rock fragments lie close to the surface, a cycle of freezing and thawing occurs, causing growth of ice under the rock. Continued thickening of the ice layer heaves the rock upward and it rises to the surface. Large angular blocks of rock in an accumulation known as felsenmeer (German for "rock sea") are a conspicuous display of frost action above treeline in the high mountains of Colorado.

Frost action moves the rocks both sideways and upward, and the heaved rocks may form bands, circles, nets, and polygons called patterned ground. A popular stopping point along Trail Ridge Road in Rocky Mountain National Park is Rock Cut where a trail leading from the road provides easy access through felsenmeer and patterned ground.

Large rocks standing on end (felsenmeer) from periglacial processes on Trail Ridge Road

During the summer season in the high country of Colorado, water is unable to percolate into an impervious layer of frozen ground below the surface. As a result, an active layer of soil becomes supersaturated and flows. This process, called solifluction, can occur on slopes as gentle as 2 or 3 degrees. Where there is a well-developed mat of vegetation, a solifluction sheet may move downward in a series of well-defined lobes and form terrace-like features. Rates of downslope movement vary depending on local conditions, but rates of about two inches per year are typical for Colorado's Front Range.

Solifluction terracing near Keystone. Note the contrast between the smooth slope on the right and the terraced mountainside on the left.

Frost action creates this interesting patterned ground on Trail Ridge Road

Rock Glaciers

Colorado has some of the most spectacular rock glaciers in the world. Active and inactive rock glaciers are common in many areas of the Colorado Rockies. Rock glaciers' over-all shape resemble ice glaciers but look like a field of rocks with no ice visible. Rock glaciers may contain ice at their cores, or may simply have ice between the rocks that deforms, allowing the glacier to flow.

Rock glaciers may be a mile or more long and have steep, unstable fronts, which may exceed 300 feet in height and stand at angles approaching 45 degrees. Large boulders and smaller stones typically form the surface, which commonly has arcuate ridges and lobes. Rock fragments within the rock glacier are smaller on average than those on the surface. Movement rates for active rock glaciers are variable, ranging from less than eight inches per year in the Front and Sawatch Ranges to as much as twenty-four inches per year in the Elk Mountains.

Rock glacier cascading off
Mount Sopris

Toe of Mount Sopris rock glacier

Terminus of rock glacier: Rock glacier on east side of East Maroon Peak

Lobate rock glaciers: Rock glaciers on west side of Tenmile Range, Copper Mountain in distance

Arcuate ridges: Rock glacier in Silver Basin south of Silverton

High, steep front of rock glacier on east side of East Maroon Peak

Sackungen

A sackung is a sort of fault that apparently doesn't have anything to do with tectonic movements. It is part of post-glaciation collapse of a mountain. When glaciers carved deep valleys and filled them with a half-mile or more of ice, the mountain valleys were so excessively steepened that when the glaciers melted, there was nothing left to support the walls of the valleys. The mountains bulged out into the valleys under their own weight and the force of gravity.

Small faults accommodated this movement. The scarp that forms is called a sackung. Sackungen are found along ridge crests in many glaciated areas of Colorado. The scarps nearly always face uphill. When sackungen are on both sides of a ridge, they can form a crestal graben. Trail Ridge Road in Rocky Mountain National Park follows a crestal graben for a while. It is possible that movements on sackungen might be triggered by large earthquakes on nearby faults.

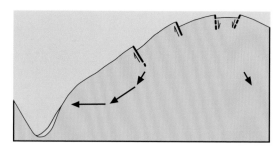

Cross section of spreading mountain: A mountain moves out into the oversteepened valley and causes its crest to collapse inward forming uphill-facing scarps called sackungen.

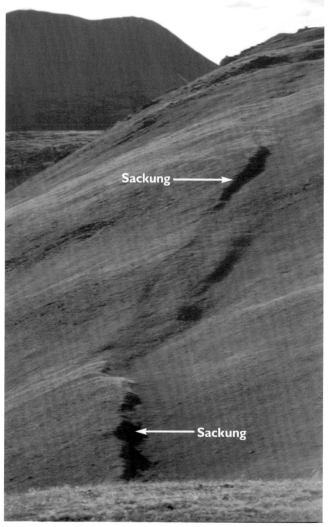

A sackung visible from Trail Ridge Road

CGS geologist standing at top of sackung scarp on Keystone Mountain. Arrows point to top and bottom of scarp.

Quaternary Wind Deposits

Colorado's dry and windy climate during the Quaternary Period created widespread deposits of wind-blown "eolian" material. Colorado has examples of most of the types of eolian deposits and landforms: loess, parabolic dunes, star dunes, barchan dunes, transverse dunes, sand sheets, reversing dunes, crescentic dunes, and blowout dunes.

Wind-blown deposits are widespread in Colorado. Deposits are located in the southwest in the Paradox Valley, in the northwest near the Little Snake River, in north-central Colorado along the eastern edge of North Park, and in the Great Sand Dunes National Park in the San Luis Valley of south-central Colorado. The most extensive eolian deposits are found east of the Rocky Mountains on Colorado's high plains where more than 30,000 square miles are covered by wind-laid deposits of loess (70 percent) and sand (30 percent).

Loess is a wind-blown deposit of chiefly silt particles that blankets the landscape. Loess deposits reach thicknesses of more than 150 feet near Beecher Island, although thicknesses of six to fifteen feet are more common. Thick loess was deposited during the last glacial period, when the most extensive deposits of wind-blown sand in North America were deposited throughout the central plains.

The South Platte and Wray dune fields of northeastern Colorado are two of the state's largest dune fields. Together, they cover nearly 5,000 square miles. Here, the winds during the last glacial period blew predominantly from the northwest and caused the dunes to migrate in a southeasterly direction. Parabolic dunes predominate in this part of Colorado. Some of these dunes were re-activated during the "dust bowl" of the 1930s.

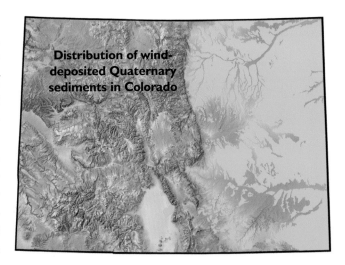

Distribution of wind-deposited Quaternary sediments in Colorado

Sixty-foot-thick loess deposit near Wray

A barefoot walk on a windy afternoon in the Great Sand Dunes National Park gives a real sense of the living nature of the dunes. You can actually see and feel the wind moving the sand grains along and above the surface. Climbing up the steep slope of a dune, you quickly learn why it is called a "slip face." Not only do your feet slip back, but also you create little landslides of sand on these steep faces that are at, or near, their angle of repose.

Dominant Winter Wind Direction

Transverse and Barchan Dunes

Great Sand Dunes

Colorado's most famous dunes are found in Great Sand Dunes National Park where the mounds of sand grow to the height of a seventy-story building, twenty stories higher than the tallest building in downtown Denver. The dunes are concentrated in the crook of the Sangre de Cristo Range.

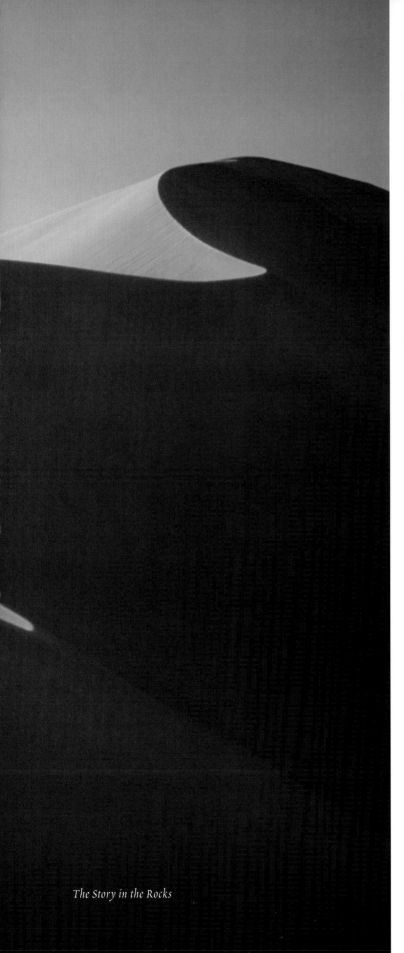

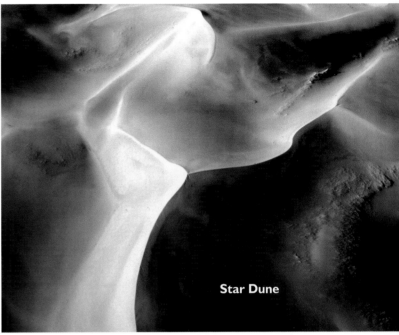

Star Dune

The wind blows sand grains up the gentle side of a dune and they slip or fall down the steep slip face. This process causes the dune to migrate in the direction the wind is blowing. It also creates large-scale cross bedding. When a trench is cut through a sand dune, you will see that the layers formed by the sand are steep and roughly parallel to the slip face. Observing such large-scale cross bedding in ancient rocks is one clue that the rock was originally an ancient sand dune.

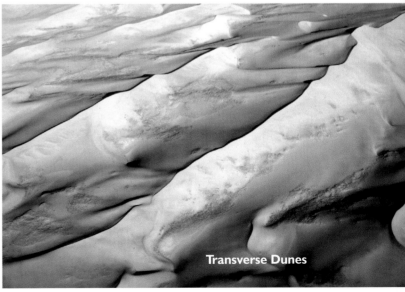

Transverse Dunes

Parabolic dunes on the High Plains

Two smaller dune fields (light-colored patches) stacked up against the Medicine Bow Mountains on the east side of North Park are visible in this digital elevation model and satellite image.

Landforms of Colorado's Landscape

Arches

Looking at a clear blue sky or a pale sliver of the Moon framed by a soaring natural stone arch can be a deeply moving experience. Maybe it is our impression of rocks as solid, dependable, and strong forces that makes these airy, oddly delicate-looking structures fascinating. Maybe we are awed by a realization that even rock is not permanent and will yield, a few grains at a time, to the patient, persistent ministrations of wind and water.

The most striking arches are in massive sandstones on the Colorado Plateau. In Rattlesnake Canyon west of Grand Junction, there are thirty-five arches carved in Mesozoic sandstones. Locals point with pride to this assemblage,

calling it "Colorado's Parade of Arches." Although it is difficult to prove definitively, some authorities claim that it is the second largest concentration of natural arches in the world. World record holder or not, the arches are a beautiful and intriguing addition to the area, as are the arches in parts of central Colorado formed in limestone, granite, and sandstone.

Shelf Canyon Window, carved from Precambrian granitic rocks south of Cripple Creek

One of thirty-five arches in the Rattlesnake Canyon area, west of Grand Junction

93

The Donut, a window in Precambrian pegmatite in Bear Creek Canyon

Spanish Peaks Window, an irregular window in one of the Tertiary igneous dikes radiating from West Spanish Peak

One of the more spectacular arches in Rattlesnake Canyon, arches in this area are carved in Jurassic sandstone

An arch in Mississippian limestone above No Name, Colorado. Note the several holes dissolved out of the limestone above the arch.

MESSAGES IN STONE

Arch near Monument (Elephant Rock) carved in Tertiary sandstone

Caves

The caves that dot Colorado's landscape are no less fascinating than the grandeur of an arch. They range from impressive limestone caverns to modest clay, shelter, and ice caves. Speleologists continue to catalogue Colorado's more than 265 caves and have compiled an impressive list of types including epigenic caves, alpine caves, hypogenic caves, granite caves, ice caves, fault caves, evaporite caves, purgatory caves, clay caves, lost caves, and mine caves.

There is evidence of early human habitation in some caves, and rare animal fossils in others. Paleontologists recently discovered a rich assortment of Pleistocene fossils in Porcupine Cave in South Park including the bones of the oldest cheetahs and ferrets in the world. Natural caves were used by early miners who hoped to reduce blasting and digging in their search for mineral riches. Ninety percent of the tunnels of the American Nettie Mine above Ouray are reputed to have begun as natural caverns.

The largest caves are dissolved from limestone, usually Mississippian Leadville Limestone that is thick, pure, and extensive. The largest, Groaning Cave, is more than eleven miles long. Caves in limestone generally form in one of two ways. Epigenic caves are dissolved by acidic, carbon-dioxide-enriched water moving from the surface down into the limestone layers and are found in the alpine areas of the state. Hypogenic caves are dissolved by carbon-dioxide-enriched (and sometimes sulfate-enriched) waters moving up into the limestone layers from below.

Breezeway Cave, note stalactites precipitating from roof fracture, an excellent pathway for fluid migration

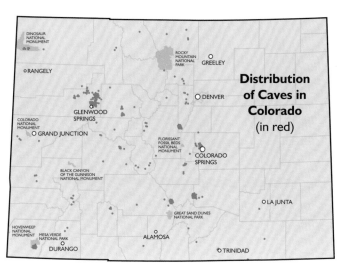

Distribution of Caves in Colorado (in red)

DINOSAUR NATIONAL MONUMENT

RANGELY

ROCKY MOUNTAIN NATIONAL PARK

GREELEY

DENVER

GLENWOOD SPRINGS

COLORADO NATIONAL MONUMENT

GRAND JUNCTION

FLORISSANT FOSSIL BEDS NATIONAL MONUMENT

COLORADO SPRINGS

BLACK CANYON OF THE GUNNISON NATIONAL MONUMENT

LA JUNTA

GREAT SAND DUNES NATIONAL PARK

HOVENWEEP NATIONAL MONUMENT

MESA VERDE NATIONAL PARK

ALAMOSA

DURANGO

TRINIDAD

Geologist points out ribbon stalactites in Cave of the Winds

Two commercial caves, Glenwood Caverns near Glenwood Springs, and Cave of the Winds near Manitou Springs, have been thrilling visitors since the 1880s. Both are dissolved from carbonate rock that is hundreds of millions of years old. Glenwood Caverns, once known as Fairy Caves, is Colorado's second largest cave. A newly opened area allows the public to view formations, but air lock chambers have been installed to preserve the humid conditions that enhance the rich colors locked in the "ribbons" of cave material.

Cave of the Winds began to form about 50 million years ago when the rock fractured during the Laramide mountain building event. About 4.5 million years ago, water rich with carbon dioxide moved upward along the cracks and began dissolving the rock. This continued for

about 3 million years until the present cave system was created. The same type of carbon dioxide water bubbles to the surface around the town of Manitou Springs today and suggests that new caves are currently being sculpted.

Although the limestone caves are most impressive, there are many other types of caves in the state including ice caves at the base of glaciers and shelter caves, which are natural cavities large enough to permit human entry but not extending into total darkness. The best-known shelter caves are in Mesa Verde National Park. Clay caves form where rivers erode banks composed of decomposed shale. More than 100 clay caves have recently been documented between Montrose and Grand Junction. One of them is 2,000 feet long. A cave along Highway 6 west of Golden, is assumed to have opened by fault movement. Purgatory caves form in igneous or metamorphic rock where slot canyons are eroded and then covered by rocks falling into the crevasses. Eighteen "lost" caves are listed in old Colorado records, but have not been rediscovered.

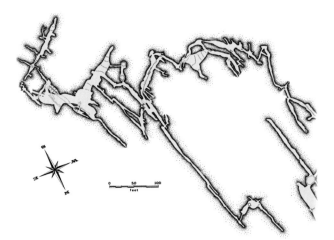
Note the rectilinear nature of the passageways in Cave of the Winds. Geologists believe that systematic fracturing of the limestone during the Laramide mountain building created fractures for CO_2-bearing waters to move along and dissolve the rectilinear passageways.

Glenwood Caverns, cave "bacon"

Ice cave located at the base of Rowe Glacier

Spelunker begins descent into Hurricane Cave, the world's deepest granite cave. Torrential glacial meltwaters carved Hurricane Cave in granite along a fault on the slopes of Pike's Peak. It is a half-mile-long and 770 feet deep. Although Pike's Peak is in the middle of a heavily populated area, the exact location of entrances to the cave—which is considered extremely dangerous—is a closely guarded secret.

Clay cave caused by water creating "pipes" through the decomposed shales near Montrose

Entrance to a cave in evaporite strata near Carbondale

Plateaus, Mesas, Buttes, Chimneys, Teepees, and Hoodoos

These landforms have one thing in common, they are all composed of nearly horizontal layers that are capped by a resistant rock layer, usually sandstone or volcanic rock, which protects more easily erodible strata beneath. These features form a continuum from plateaus (measured in thousands of square miles) down through hoodoos (measured in thousands of square inches).

The exact size at which a feature is called a mesa or a plateau, a mesa or a butte, or a butte or a chimney is not precisely defined. It is sort of "when you see it, you'll know it." Western Colorado is part of the continent-scale Colorado Plateau province. This 100,000-square-mile feature is broken into smaller plateaus such as Colorado's 2,500 square-mile Uncompahgre Plateau.

Above: Mount Garfield east of Grand Junction along I-70 is an excellent example of how resistant sandstones protect underlying weak shales. Note that the sandstones are bounded by vertical cliffs, whereas the gray shales form more gentle slopes.

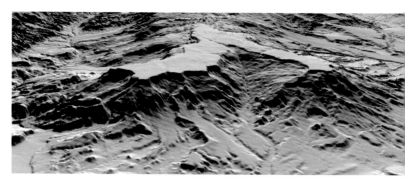

This basalt-capped mesa, Grand Mesa, east of Grand Junction is promoted as the world's largest flat-topped mountain. It measures 380,000 acres, has more than 300 lakes, and is 10,500 feet above sea level. The basalt was originally poured out onto a relatively flat plain about 8 million years ago. The weak sedimentary rocks below the volcanic cap are eroding rapidly and are highly susceptible to landslides.

Uncompahgre Plateau south of Grand Junction is made of resistant Jurassic sandstones that protect red Triassic shales over 2,500 square miles.

The resistant Pawnee Buttes were left behind as erosion caused the High Plains escarpment's retreat to the east. The escarpment marks the boundary between the Colorado Piedmont and the High Plains subprovinces of the Great Plains physiographic province.

Pawnee Buttes

High Plains Escarpment

Pawnee Buttes, northwest of Sterling, are capped by resistant conglomerate of the Ogallala Formation shown in the foreground.

Lizard Head: Many (including some geologists) have thought this spire in the San Juan Mountains was a volcanic neck. Rather, it is made of horizontal layers of tuff protected by a horizontal layer of welded tuff.

Cretaceous sandstones protect Cretaceous shales in Chimney Rock on the right and the mesa on the left. Chimney Rock was part of the mesa until the upper sandstones were eroded through and the shales below were removed.

The Teepees: These conical hills of Cretaceous shales east of Pueblo are capped by patches of resistant, fossiliferous limestone.

Earth Pillars: These boulders along I-70 protect the underlying soft material from the forces of erosion and create the erosional pillars or hoodoos.

Hoodoos: These pillars of Cretaceous sandstone in Monument Park are protected by thin, more strongly cemented layers of sandstone.

River Valleys

River valleys in Colorado vary widely in their depth and shape from the narrow, sheer walls of the Black Canyon of the Gunnison to the broad valley of the Colorado River in the Grand Valley near Grand Junction. Meandering rivers are most commonly found on the plains and plateaus of Colorado. However, such streams do occur in some mountainous areas, usually indicating that the pre-glacial stream gradient was flattened by glaciation.

Above: East River, near Mount Crested Butte, is atypical for mountain streams. A large glacier occupied the valley and modified the stream gradient, making the valley bottom much flatter.

Within ten miles, the Platte River changes from a mountain stream with steep walls and rushing waters (right) to a languid stream (below) with the reputation that it is "too thin to plow but too thick to drink." In the mountains, the valley is about one and a half miles wide and 250 feet deep. On the plains, the Platte River valley is still about 250 feet deep but seven miles wide.

Black Canyon of the Gunnison, cut into Proterozoic igneous and metamorphic rocks by the Gunnison River, is as much as a half-mile deep. No other canyon in North America combines the narrow opening, sheer walls, and startling depths offered by this canyon.

The rate of downcutting over the past 2 million years is estimated to be about one inch per hundred years. Quaternary faulting just south of the canyon indicates that the rapid downcutting is caused by recent uplift.

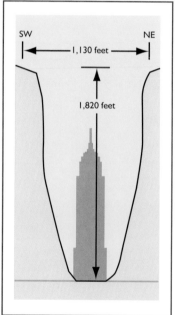

Profile of Black Canyon of the Gunnison at Chasm Overlook, scaled in feet; silhouette of Empire State Building is at same scale

Hogbacks and Flatirons

Whereas mesas form from resistant, flat-lying strata, hogbacks form where sedimentary strata are turned up on end by a large basement uplift. The actual hogback ridge forms where resistant strata are surrounded by erodible rock that leaves the resistant layer relatively isolated as a hogback ridge.

Hogbacks are somewhat separated from the mountain front. Flatirons are most commonly resting against the mountain front. Both have strata that have been turned up by an uplifted feature. Flatirons form from a combination of the dipping strata and differential erosion that creates shapes that resemble the bottom of a hand iron.

The Dakota Hogback west of Denver is notable because it is the start of the Rockies for those coming from the east. It is held up by resistant Cretaceous sandstone.

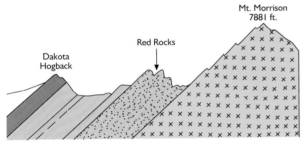

Cross section west of Denver illustrating how the tilting of layers and differential erosion create the Dakota Hogback

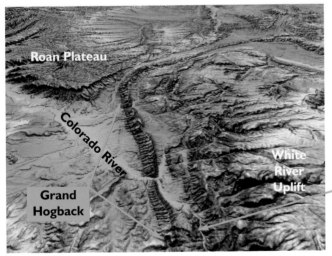

Strata in the Grand Hogback near Glenwood Springs were turned up when the White River basement block was uplifted during Laramide mountain building.

Outcrop of dipping strata of the Grand Hogback along I-70 west of Glenwood Springs

South of Durango, this monocline was created when sedimentary strata were upturned by the large San Juan basement uplift to the north during Laramide mountain building.

Coloradoans enjoy taking their flatland friends for a thrilling car ride on Skyline Drive, a hogback above Cañon City.

The steep angle of the Pennsylvanian beds combine with the forces of erosion to create the famous Boulder Flatirons. The flatirons were tilted when the Front Range block was pushed up during the Laramide mountain building. Try to find the geologist for scale.

Anatomy of the Mountain Front West of Denver

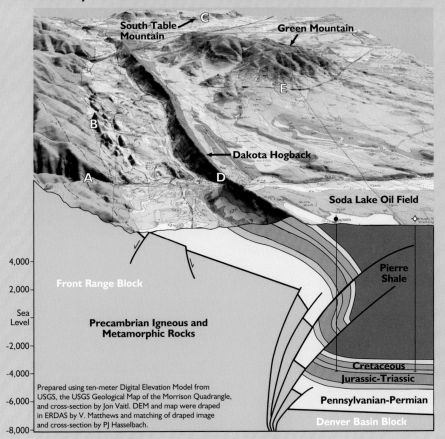

South Table Mountain

Green Mountain

C

E

B

← Dakota Hogback

A

D

Soda Lake Oil Field

Front Range Block

4,000–
2,000–
Sea Level
–2,000
–4,000
–6,000
–8,000

Pierre Shale

Precambrian Igneous and Metamorphic Rocks

Prepared using ten-meter Digital Elevation Model from USGS, the USGS Geological Map of the Morrison Quadrangle, and cross-section by Jon Vaitl. DEM and map were draped in ERDAS by V. Matthews and matching of draped image and cross-section by PJ Hasselbach.

Cretaceous
Jurassic-Triassic
Pennsylvanian-Permian
Denver Basin Block

The large image shows the geology of the Morrison seven and one-half-degree quadrangle west of Denver. Note that the layers of sedimentary rock in Green Mountain are horizontal, but the layers near the uplifted Precambrian rocks of the mountain front are bent up by the movement on the mountain block.

The Front Range has been uplifted a minimum of 21,000 feet based on the depth of Precambrian rocks in the Denver Basin (7,000 feet below sea level) and the altitude of Precambrian rocks in Mount Evans (14,264 feet above sea level). The cross section is constructed from the rocks exposed at the surface, the rocks encountered in the deep wells that are in the area, and information from seismic studies.

Precambrian rocks near mouth of Bear Creek Canyon

Left: Tilted Paleozoic sedimentary rocks at Red Rocks. Precambrian rocks underlie the grassy slope to the left

Right: South Table Mountain contains nearly horizontal layers of Cretaceous and Tertiary sedimentary rock capped by basalt-like flows

Left: Inclined layers of Cretaceous sedimentary rocks in the Dakota Hogback along CO-74

Right: Nearly horizontal layers of Cretaceous sedimentary rocks in Green Mountain

Landforms of Colorado's Landscape

This site along I-70 and the Colorado River illustrates why the deposit is called an alluvial fan. This fan is fed by two canyons—one large, one small. The water, when confined in the narrow canyons, has tremendous power to erode and transport materials. When the water is no longer confined by the canyon it fans out and loses its power, dropping its load of sediment.

Great Sand Dunes

Fans

Huge alluvial fans bank up against Sierra Blanca. These fans are 2,000 feet high and probably the largest fans in Colorado. The fans reflect the active uplift of Blanca Massif (14,345 feet). The Great Sand Dunes are in the background in the bend of the range.

Quaternary Alluvial Fans

Alluvial fans are prominent in many parts of Colorado but are most spectacular along the west flank of the Sangre de Cristo Range. Alluvial fans form where debris-laden mountain streams emerge from steep, narrow canyons onto wide valley floors. Indeed, in engineering geology they are commonly referred to as debris fans in order to emphasize the debris and mudflow events that can dominate fan deposition and that are potentially very lethal and damaging. Unfortunately, because fans commonly have nice views, property owners sometime choose to build on them, which is not a good idea because they also are the location for many active geologic hazards.

Construction of homes with a view on a fan in East Vail

106

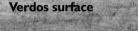

Rocky Flats surface

Verdos surface

These are two of the more prominent, flat terraces identified along the eastern flank of the Front Range. The Verdos surface is about 600,000 years old and is 100 feet lower than the older Rocky Flats surface but 250 feet above the modern stream bottom. This view is from CO-93 between Boulder and Golden.

Quaternary Terraces and Pediments

Streams that are in balance so that they are no longer cutting down but are predominantly cutting sideways create broad, flat valleys. If the river is put out of balance by mountain uplift or dramatic climate change, the river may begin cutting down to a new grade, leaving the old broad valley floors high and dry as terraces. Quaternary Colorado terraces are associated with glacial cycles and are capped by surficial deposits of coarse sands and gravels. The deposits, although correlated with glacial cycles, are not necessarily glacial in origin. Pediments are surfaces that have been planed off by lateral erosion and then mantled with a veneer of alluvium.

A terrace is a broad, nearly flat landform bounded by steep slopes on either side. Terraces can generally be linked to a specific canyon mouth at the mountain front. In some valleys, several terrace levels can be seen along the sides of streams and rivers. The upper Arkansas River Valley, near Buena Vista, is noted for the step-like terraces that flank both sides of the river. Along the Front Range, five major terrace levels are recognized and named.

Terrace deposits are differentiated and assigned ages primarily on the basis of height above stream level. In some areas, terraces can be dated by their position relative to terrace deposits containing volcanic ash, which has been radiometrically dated. Some of the terraces in the Yampa Valley, for instance, are 620,000 years old.

This is typical coarse gravel found capping terraces such as the one below in western Colorado

This large terrace near Rifle was eroded before the Colorado River had cut down to its present level. It is capped with a deposit of coarse gravel.

The inclined layers are planed-off. Cretaceous sedimentary rocks are unconformably overlain by horizontal, Quaternary (Verdos) pediment sediments

Quaternary Volcanoes

The oldest known Quaternary volcanism in the state occurred about 1.5 million years ago in the Roaring Fork River Valley, ten miles north-northwest of Aspen. Magma rose along the Crystal River fault and broke out on the side of the valley producing a cinder cone and small flows totaling about 500 feet in maximum thickness. Along Rock Creek, just northeast of McCoy, volcanism produced two cinder cones and a basalt flow, which spread out on an old Rock Creek floodplain. The flow has been radiometrically dated at approximately 640,000 years old. More recent eruptions have been recorded near the junction of the Colorado and Eagle rivers where volcanic flows overlie quite modern topography. West of the junction, high-level volcanism of uncertain age produced a cinder cone (known as Willow Peak) and a basalt flow that descended the upper part of a small valley tributary to the Colorado River.

The basalt pushed the Eagle River southward

The most recent volcanism known in Colorado appears near the town of Dotsero. Using a charcoal sample recovered from a tree that had been buried by the falling ash, the Dotsero volcano and associated flow have been radiometrically dated as 4,150 years old.

Basalt worked its way to the surface 4,150 years ago and flowed from a vent in the hills of evaporite down to the Eagle River, probably damming it for a short time. The flow is outlined with yellow dashed lines. Directions of flow can be determined by the orientation of the lobes and the orientation of the curvilinear furrows and ridges.

These layers are composed of basaltic cinders that built the cone over the vent that fed the Dotsero basalt flow. Mining has removed most of the cinder cone.

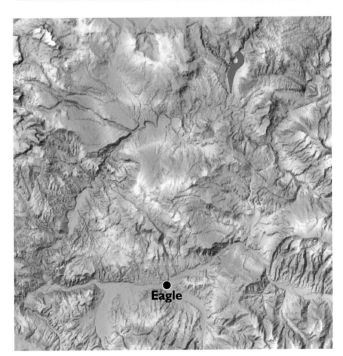

Quaternary basalts in west central Colorado

Quaternary Basalts in San Luis Valley

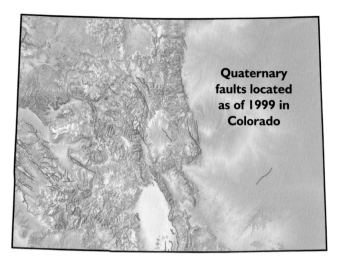

Quaternary faults located as of 1999 in Colorado

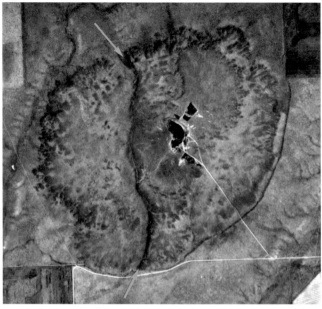

Aerial photograph of Quaternary fault scarp in the Quaternary Mesita cinder cone in the San Luis Valley

Quaternary Faulting

Faults have broken Quaternary deposits in many basins in Colorado. According to a recent inventory by the Colorado Geological Survey, there are at least ninety faults in Colorado that moved during the Quaternary Period. Eight of these faults have documented movement in the past 15,000 years. Many, but not all, of these younger faults are associated with the Rio Grande Rift.

Faceted spurs on the flanks of the Sangre de Cristo Range in the San Luis Valley mark the bounding fault and are a testament to the recent uplift and faulting in this area.

CHAPTER 4

Humans and Geology
Natural Riches: Colorado's Economic Geology

Gold nugget

Colorado has a great variety of mineral resources that society, both modern and prehistoric, has found useful or necessary for survival. Indigenous people mined clay and sand to use in house construction as well as to make pots and decorative items. Obsidian, chert, chalcedony, and other forms of silica were quarried and used to manufacture tools and weapons 13,000 years ago. Hard, crystalline igneous and metamorphic rocks were used for grinding stones. Tools and weapons made from Colorado resources were transported to, and traded in, neighboring states. The economic importance of Colorado's mineral and mineral fuel resources continues today. The total value of production in 2000 was $4.5 billion.

Colorado's most famous mines, and the cornerstone of the state's pioneer history, are those that produced precious metals. When prospectors discovered gold in the gravel of the Cherry Creek and South Platte Rivers in 1858, the rush was on. The pursuit of gold is responsible not only for the development of many mines, but also for settlement of many new cities including Denver. Later discoveries resulted in metal mines throughout the mountainous areas of the state where gold, silver, lead, zinc, molybdenum, copper, and tungsten were mined. More than 770 minerals have been catalogued in Colorado.

Energy resources of oil, natural gas, oil shale, uranium, and coal are abundant. Crude oil and natural gas are produced from more than 55,000 wells in more than half the counties of the state. Colorado coal resources are large, the quality is very good, and the coal industry in Colorado is enjoying resurgence. There are also enormous deposits of oil shale in Western Colorado.

Industrial minerals and construction materials, the unheralded, working-class plodders of the mineral resources industry, are literally the building blocks of our society. Colorado is fortunate to possess abundant resources of several commodities in this category.

Remnants of Colorado's gold mining boom at Rustler Gulch

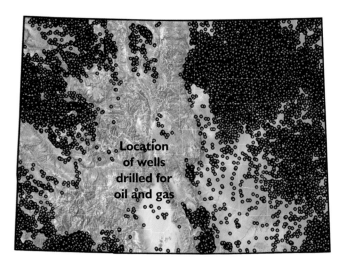

Location
of wells
drilled for
oil and gas

NATURAL GAS, OIL, AND OIL SHALE

In the spring of 1862, just two years after Colonel Drake completed the world's first producing oil well near Titusville, Pennsylvania, the second oil field in the United States was established near Cañon City, Colorado. In 1876, oil was found in fractures and fissures in the Pierre Shale near the town of Florence. The Florence and Cañon City fields have produced more than 15 million barrels of oil from 1,300 wells since then. Florence is an unusual field because the accumulation is in a syncline rather than an anticline.

Contrary to popular belief, oil does not occur in pools or caverns underground. Rather, it fills the many small pore spaces between the sand grains such as in this Permian sandstone from western Colorado wherein gray streaks are oil-saturated layers.

Oil well near Longmont

The first gas production in Colorado was established in 1890 in northwestern Colorado's Piceance Basin. Northwest Colorado is also the home of Rangely Oil Field, the largest oil field in the Rocky Mountain region. Oil production was first established on this huge anticline in 1902. Large structures in the Rocky Mountains, such as Rangely, are affectionately referred to as "sheepherder anticlines" because they are so large and so obvious that even a geologically untrained sheepherder can recognize their potential.

Cumulative production from Rangely Field at the end of 1999 was 845 million barrels of oil and 761 billion cubic feet of gas. During its heyday in the early 1950s, Rangely produced in excess of 62,000 barrels of oil per day. Today it is nearing depletion and produces only 18,500 barrels daily.

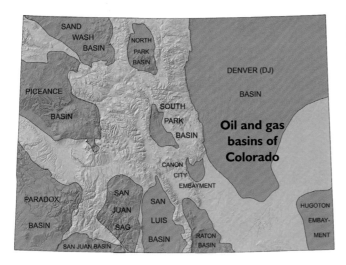

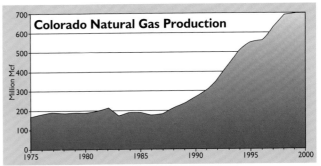

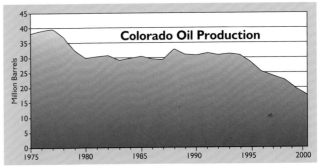

Colorado oil and gas production from 1975 through 2000

Historically, thirty-six of Colorado's sixty-four counties have produced oil, and thirty-nine counties have produced natural gas. Cumulative production from the nearly 1,400 fields that have been discovered in Colorado now stands at about 1.83 billion barrels of oil and 10 trillion cubic feet of gas. In 2000, Colorado ranked eleventh in the United States in oil reserves, and fifth in gas reserves.

In recent years, natural gas production has been on the increase in Colorado while crude oil production has declined. Colorado's four principal oil and gas producing regions are the Piceance Basin and Sand Wash Basin region, the Paradox Basin and San Juan Basin region, the Las Animas Arch region, and the Denver Basin region.

On hot, summer days, crude oil naturally seeps out of Cretaceous sandstone in an outcrop at the interchange of C-470 and US-285.

If traditional fossil fuels run out in the twenty-first century, some think that Colorado's oil shale reserves might produce 600 to 800 billion barrels of oil. The name "oil shale" is really a misnomer: it's not really shale and doesn't contain oil. It is actually organic marlstone. Marlstone is a mixture of clay and calcium carbonate. The organic material is kerogen, a precursor to oil. In order to yield oil, marlstone must be mined, crushed, and retorted at 900 degrees Fahrenheit. The key to production is the development of an efficient, economic technology. In addition to the mining and processing technical problems, there are major obstacles relating to the amount of water needed and the amount of waste produced.

In the late 1970s, virtually every major oil company was actively trying to produce Colorado's oil shale economically—all failed. Despite a century of interest in this resource, energy companies have not yet developed cost-effective methods of retrieving the oil, and the feasibility of developing this resource in the future is uncertain. However, as world supplies of oil near peak production, major companies are once again eyeing this enormous resource.

This drill rig near Rifle is part of the boom in clean-burning, natural gas development in Colorado. Looming on the cliffs above are outcrops of oil shale on the Roan Plateau, a reminder of the boom and bust of the oil shale industry in Colorado in the late 1970s and early 1980s.

COAL

Coal is one of Colorado's most widespread mineral and energy resources, underlying 29,600 square miles or 28 percent of land in the state. The majority of coal was formed approximately 80 to 65 million years ago in the swamps of the late Cretaceous and early Tertiary periods.

Coal has been commercially mined for at least 140 years in the state. Some of the earliest coal mining began in the 1860s in the Denver Basin north and west of Denver. Early coal mining also began around the turn of the twentieth century in the Raton Basin west of Trinidad. These two basins supplied 38 percent of all coal ever produced in Colorado. Although both basins have been inactive for several decades, a new mine has recently started up in the Raton Basin. Today, most of the coal production comes from the northwest part of the state.

Coal mined underground near Paonia is piled at the surface after washing. Here it waits to be loaded in train cars and begin the long journey to eastern markets.

An open-pit coal mine near Craig

Mine geologist measures the thickness of a coal seam in an open-pit mine near Hayden

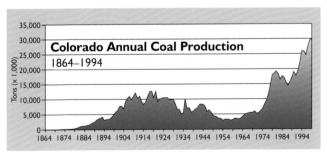

Colorado Annual Coal Production 1864–1994

Coal on the flanks of the pit is removed using giant auger bits near Steamboat Springs

Colorado coal is especially valuable because of its relatively low content of sulfur, ash, and mercury. If present, these constituents pose environmental and residue problems when coal is burned. In order to comply with the environmental standards set by the Clean Air Act, electric generating plants in the eastern U.S. have found it necessary to purchase the clean, but more expensive, Colorado coal to mix with their sulfur-laden, local coal.

Approximately 45 percent of the state's coal production is burned in Colorado power plants, while 49 percent is shipped to other states. The remaining percentage is shipped to foreign countries. Within the state, power plants have been built near the major coal mines in northwest, west, and southwest Colorado, as well as along the Front Range from the Wyoming border to Colorado Springs.

The Energy Information Administration (EIA) estimates that Colorado has approximately 16.5 billion tons of coal reserves. Colorado is well situated to continue to provide high-quality, clean-burning coal to its citizens as well as to users in other states and countries.

COALBED METHANE

Coalbed methane comes from biologic decay of peat (swamp gas) and/or methane generated during burial and heating of the organic material as it converts into bituminous coal. As far back as the 1700s in England, catastrophic methane explosions plagued underground coal mining. The colorless, odorless, highly explosive gas is capable of spontaneous combustion within a closed space and has been a menace to coal miners throughout the ensuing centuries. Today, however, coalbed methane is a valuable source of energy and helps mine safety.

Today, the San Juan Basin of Colorado and New Mexico is the most prolific coalbed methane-producing basin in the world. Coalbed methane accounted for 54.5 percent of Colorado's total gas production in 1998. These productive coalbed methane areas are in only seven of the state's sixty-four counties. That number is bound to increase in the future as technological advances and geologic understanding of coalbed methane continues.

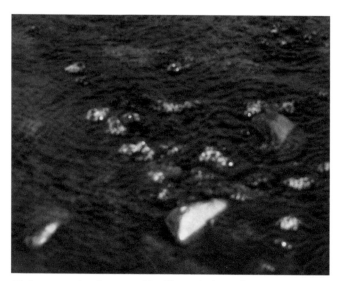

Methane escaping from a coal bed beneath the Little Snake River bubbles up through the water in northern Colorado

METAL DEPOSITS

The early history of the State of Colorado runs parallel to the history of metal mining and is directly tied to the first significant and documented discovery of gold in the summer of 1858. A party of prospectors discovered placer gold in stream gravels at what is now downtown Denver, near the confluence of the Cherry Creek and South Platte Rivers (Auraria). This discovery led to the first Colorado gold rush. "Pike's Peak or Bust" was emblazoned on many of the ox-drawn wagons carrying optimistic prospectors.

In the Denver area, gold deposits proved to be small and were quickly depleted. The spirits of disappointed prospectors again soared when subsequent searching in Clear Creek Canyon led to the discovery of the rich vein and placer gold deposits at Idaho Springs and Central City. By the time Colorado became a state in 1876, many mining districts and new cities had sprung up throughout the area.

At various times throughout its history, Colorado has been the leading U.S. producer of gold, silver, molybdenum, lead, zinc, uranium, and tungsten. In 2000, it ranked second in molybdenum production, behind Arizona, and seventh among the states in gold production. Other metals that have been mined in Colorado include copper, tin, vanadium, iron, beryllium, lithium, rare earth elements, thorium, tantalum, and manganese.

THE COLORADO MINERAL BELT

Most of the significant metal deposits in Colorado, with some notable exceptions, are located within a northeast–southwest trending zone ten to sixty miles wide. Extending from the Front Range just north of Boulder to the La Plata Mountains northwest of Durango, this zone is known as the Colorado Mineral Belt. The famous mining districts and towns of Aspen, Telluride, Silverton, Leadville, Breckenridge, Fairplay, Crested Butte, Central City, Jamestown, Georgetown, and many, many smaller towns, ghost towns, and camps are located in this relatively narrow but productive belt.

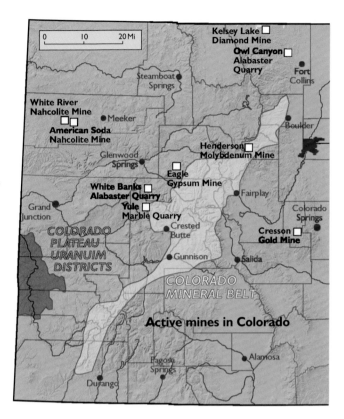

Active mines in Colorado

The precious-metal (gold and silver) and base-metal (lead, zinc, and copper) deposits of the mineral belt were created by igneous events (intrusion and volcanic eruptions), most of which occurred between 75 and 20 million years ago.

VEIN DEPOSITS

A large part of Colorado metal production, especially in the past, came from vein deposits. The veins are tabular or lenticular bodies that range from less than an inch to nearly 100 feet in width, and from tens of feet to several miles in length.

Vein formation begins with the creation of fissures (open-space cavities) formed as a response to movement of the rock along faults. Circulating hot water containing dissolved metals, silica, carbon dioxide, and sulfur compounds (hydrothermal solutions) flowed through the fissures and deposited the minerals as the solutions became over-saturated.

Although precious and base metals are the main commodities mined from hydrothermal veins in the Colorado Mineral Belt, other important metals have also been extracted from them. In the early twentieth century around the time of World War I, the Boulder area produced more tungsten than any other region in the United States.

The enormous lead-zinc-silver deposits at Leadville, Gilman, and Garfield-Monarch were also formed from hydrothermal solutions. Rather than being deposited in open-space cavities in the pre-existing rock, the ore minerals replaced other minerals, especially dolomite and calcite in carbonate sedimentary rocks forming replacement deposits. Hydrothermal solutions containing the dissolved metals ascended along faults and fractures. When the solution reached a receptive dolomite or limestone bed that was capped by an impermeable layer such as an igneous sill or a shale, the solution spread through the receptive unit, removing calcium, magnesium, and carbon dioxide and replacing it with base metal sulfides and other minerals.

CLIMAX-TYPE MOLYBDENUM DEPOSITS

The Henderson Mine near the town of Empire is the largest operating primary producer of molybdenum in North America. Molybdenum is used mainly as an alloy to strengthen steel and also as a lubricant. At Henderson, molybdenite occurs in quartz veins and veinlets within a Tertiary-age, granitic intrusion. The ore is mined underground (more than 3,000 feet below the surface) and transported by an underground conveyor belt for ten and a half miles under the Continental Divide to the ore-processing facilities in Summit County.

Henderson is geologically similar to the Climax molybdenum deposit northeast of Leadville, which is now inactive, but once was the world's largest provider of the metal. The Mount Emmons molybdenum deposit west of Crested Butte is also a world-class resource, but has not yet been developed.

PLACER GOLD

Panning for gold in creeks and rivers is placer mining on a small scale. Placer gold is gold that has been eroded from its bedrock source (probably vein deposits) and carried by water downstream and deposited with other sediment as a placer deposit. Since gold is heavier than the other rock material in the streambed, it generally sinks in the sand and gravel and is found close to the bedrock below. Flowing water acts as a natural gravity concentrator and can form immensely rich pockets of gold.

Large placer gold mines have operated in several places in Colorado. The largest of these were near Fairplay along the upper South Platte River and Breckenridge along the Blue River and French Gulch. Dredges equipped with huge buckets scooped gravel out of the riverbeds and processed it to concentrate the gold. No chemicals were necessary to extract the gold. In some places, hydraulic mining methods were used. Water under very high pressure was sprayed through large nozzles at slopes with gold-bearing material to wash the loose material down the hill where it was run through sluice boxes to collect the gold.

Gold panning instruction given on the banks of Cherry Creek in Denver, 1932

One of the last gold dredges in Colorado near Fairplay, with huge piles of processed gravel evident

Large piles of processed gravel, known as tailings, still line these rivers years after mining has ceased. Some of the sand and gravel pits near Denver along the South Platte River and Clear Creek produce small amounts of gold as a byproduct of their aggregate operations. There are many places in Colorado where you can try your hand at panning for gold.

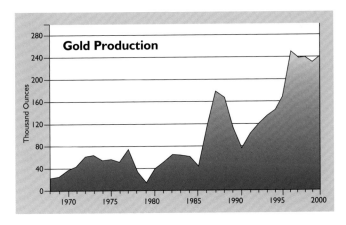

Several veins intersect in the "Glory Hole" near Central City in the heart of the Colorado Mineral Belt. After following veins underground, miners decided they could recover more of the gold by tackling it from above rather than from the underground tunnels.

METAL DEPOSITS NOT IN THE MINERAL BELT

Cripple Creek

The largest single gold mining district in Colorado by far is Cripple Creek. The mines of Cripple Creek and its neighbor, Victor, have produced nearly one half of all gold mined in the state. Over 21 million ounces of the metal have been produced in the district since its discovery in 1891. The Cresson Mine is the only active gold mine in Colorado and produced over 230,000 ounces of gold in 1999.

The Cripple Creek district lies well south of the Colorado Mineral Belt, making it an unusual deposit. The gold deposits are related to an Oligocene age (28 to 32 million years ago) intrusive center that may be the subsurface portion of an eroded volcano.

Powderhorn Titanium Deposit

The world's largest single identified resource of titanium metal is located in Gunnison County. The Powderhorn titanium deposit has not yet been mined, but it recently underwent an evaluation by a major mining company to determine the economic feasibility of a new mining operation there. Although the geological resource is very large, the extraction of titanium from its host rock is very expensive and energy intensive.

Titanium is a lightweight, silvery-gray metal used for white pigment in paint, lightweight mountain bike frames and golf clubs, and alloys used in the manufacture of space vehicles, submarines, and airplanes. It is also the new chic skin for architects to use on modern buildings. Denver is the world headquarters of the largest manufacturer of architectural titanium.

Precambrian Base and Precious Metal Deposits

Precious metals, base metals, and some tungsten have been mined in Colorado from small deposits formed in late Precambrian (Proterozoic) time. These deposits are scattered throughout the mountain region but are especially numerous in Gunnison, Saguache, Fremont, and Chaffee counties of south-central Colorado. The deposits tend to be smaller than the Laramide-age deposits.

Dotted along the rimrock of the canyon country of western Colorado are reminders of one of the greatest exploration booms the country has ever known. All you needed to participate in the rush was a geiger counter, sleeping bag, grub, and a lack of fear of heights or rattlesnakes.

Metamorphism and tectonic movement subsequent to the formation of these ancient deposits has generally made them more difficult to understand—and to mine—than the younger deposits. The Sedalia Mine near the town of Salida was the state's largest producer of copper in the late nineteenth and early twentieth centuries.

Uranium and Vanadium

The 1950s witnessed one of the most intensive treasure hunts in Colorado. Thousands of fortune hunters swarmed over every inch of the canyon lands of western Colorado searching for uranium. Many were successful.

Uranium from Colorado was used by the French physicists Madam Curie and her husband Pierre in their Nobel Prize-winning research on radioactive substances. Uranium is used to fuel nuclear power plants, create radiation for medical experiments and treatments, and to make weapons of mass destruction. Vanadium, a metal with many industrial applications, often occurs with uranium minerals.

Most deposits of uranium and vanadium in Colorado are located in the plateau region of western Colorado. These deposits usually occur in the Morrison Formation of Jurassic age and were created when groundwater carrying dissolved uranium and vanadium flowed through the porous sandstone and encountered carbon-rich, fossilized plant material. The fossilized plants acted as chemical traps, precipitating the metals out of solution. Occasionally entire fossil tree trunks, composed almost entirely of uranium minerals, have been found. The largest uranium mine in Colorado was the Schwartzwalder Mine between Boulder and Golden, which closed in 2000. Uranium mineralization at this mine occurred in veins in Precambrian metamorphic rock.

Pegmatites

Pegmatites are exceptionally coarse-grained igneous rocks occurring as vein-like tabular masses or irregularly shaped bodies in or near large granite plutons such as stocks and batholiths. Large crystals of feldspar, quartz, and mica are the dominant minerals, but many pegmatites also contain variable quantities of rare and sometimes valuable minerals. Mineral collectors like pegmatites because they can contain large, well-formed crystals and some rare minerals can be found in them. Colorado pegmatites also yield beryllium, lithium, niobium, tantalum, thorium, yttrium, cerium, and lanthanum. All of these metals have high-tech applications. Larger deposits in other areas of the world currently meet the demand for these metals, and pegmatites are currently not being mined in Colorado.

These coarse-grained granite dikes in Cotopaxi Canyon are widespread throughout Precambrian rocks in Colorado. They can contain a host of exotic elements and beautiful, gem-quality minerals.

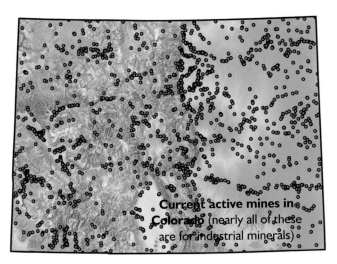

Current active mines in Colorado (nearly all of these are for industrial minerals)

INDUSTRIAL MINERALS AND CONSTRUCTION MATERIALS

We depend upon a staggering amount of industrial, or nonmetallic, mineral resources for the goods we use everyday. These raw materials are extracted and then transformed by various industrial processes into useable consumer goods. From our first cup of coffee in the morning to our commute to the magazine we read to the water we drink to the salt we add to our food to the house that gives us shelter, industrial minerals are essential to our most basic needs and most extravagant wants.

Colorado has vast resources of many minerals. Fueled by Colorado's tremendous rate of growth, the group of industrial minerals known as construction minerals is by far the most economically significant of the industrial mineral resources in Colorado. Of the construction minerals, aggregate is the most common commodity. Aggregate is sand, gravel, and crushed rock—the largest components of concrete, asphalt, and road base. It is estimated that each person in the Front Range uses about fourteen tons of aggregate per year. Only production of natural gas, oil, and coal surpasses the value of Colorado aggregate production.

Wherever there is construction, sand and gravel are needed. Colorado ranked eighth in the nation in 2002 for sand and gravel production with a little more than 45 million tons. Wherever there has been fluvial transport of sediments, such as by streams or glaciers, there is a good chance of finding sand and gravel. All along the Front Range there are sand and gravel deposits. Some of the best quality gravels are along the South Platte, Cache La Poudre, and Clear Creek Rivers. Although quality is important, the distance of transporting sand and gravel to construction projects is the single-most important factor influencing the price. It is therefore crucial to locate high-quality deposits near the places where construction is occurring.

Another material used in construction is crushed rock. Crushed rock is typically used as road base and in asphalt, but also is frequently used as decorative stone. As more and more sand and gravel deposits are rendered inaccessible by development, demand for crushed rock is increasing. Fortunately, there is an almost unlimited supply of granite and gneiss in our mountainous regions, which can be used for this purpose. Sandstone and limestone are also good sources of crushed rock, but deposits are not as large.

Most of the industrial-grade limestone and dolomite currently being mined in Colorado is from the Cretaceous strata along the Front Range, especially at La Porte and near

The gray blocks used in constructing Molly Brown's house are known as "Castle Rock Rhyolite" and are actually ash-flow tuffs of the 36-million-year-old Wall Mountain Tuff. The red trim is of Permian sandstone from Manitou Springs, and the flagstones in the sidewalk are from Permian dune sandstone near Lyons.

Portland. Limestone and dolomite are widely used by the construction industry. Crushed limestone and dolomite are used as road base or decorative stone, and large blocks of limestone and dolomite can be used as building stone. Lime is also the key ingredient in cement and is frequently used in agriculture as a soil additive to help break up clay soils, neutralize acidic soils, and as an essential plant food. Limestone has also been quarried for use in the sugar refineries of the eastern plains.

One of the most useful industrial materials is clay. There are hundreds of uses for clay, but in Colorado it is used primarily to make bricks. The Cretaceous formations are major sources of clay. The Coors Porcelain Company has mined clay for many years for use in pottery and low-temperature ceramic ware.

Stone that is cut or shaped to a specific dimension, generally for building purposes, is dimension stone. In Colorado marble, granite, sandstone, rhyolite, and travertine are used for dimension stone.

A common mistake is for geologists new to the Front Range area to think that these scars in many of the steeply dipping sedimentary layers are faults. However, they are scars left from mining thin clay layers.

The Permian Lyons sandstone of Boulder and Larimer counties has been used as dimension stone since the mid-to-late 1800s. The Aberdeen Granite of Gunnison County is one of the most widely used granites in the state. The state capitol and the grayish steps leading up to the entrances contain 283,500 cubic feet of this stone. The colorful interior walls of the capitol are of limestone (called "Beulah Marble") quarried west of Pueblo. Cotopaxi Granite of Fremont County was used in the construction of the base of Denver City Hall. Pike's Peak Granite of Douglas County and Silver Plume Granite of Clear Creek County are other notable sources of dimension stone.

The most significant travertine deposit in Colorado is located near Wellesville. Denver General Hospital, the Gates Rubber Company, and the Bus Terminal Building are just some of the buildings in the Denver Metro area constructed of Colorado travertine. The light gray, Wall Mountain Tuff of Douglas County, known in the building industry as the Castle Rock Rhyolite, has been widely used as building stone in the Front Range. One of Denver's favorite landmarks, the Trinity United Methodist Church, was built using this rock.

Gypsum, salt, and nahcolite are industrial minerals with a common origin as evaporites deposited in Permian-Pennsylvanian inland seas. The most widespread deposits are in the Paradox Valley in southwestern Colorado. The cost of shipping the evaporite commodity determines the economic viability of the deposits. Far from major transportation routes, these deposits have not been extensively mined. In contrast, gypsum deposits in the Eagle Basin near the town of Gypsum are mined to make drywall for use in building construction. This mine and its drywall plant are located adjacent to I-70 and only a few miles from a rail line, which makes shipping cheap and easy.

Nahcolite is the mineral name for sodium bicarbonate (baking soda). In the Piceance Basin, nahcolite is mined by pumping hot water down into the deposit where it dissolves the nahcolite and is pumped back to the surface where the bicarbonate is captured. Most of the baking soda is exported, but a small percentage is sold in Colorado to make snack foods.

This gypsum mine is located near, of all places—Gypsum

Construction of a plant to convert the sodium bicarbonate to sodium carbonate began in 2000. Sodium carbonate is commonly known as soda ash and is more widely used in industry. It is an important ingredient in glass and fiber optics, chemical manufacturing, and is used to remove sulfur dioxide (a major pollutant) from the flue-gas of coal-fired power plants. The northwestern Colorado deposit may contain 30 billion tons of nahcolite, which would make it the largest in the world.

Nahcolite mining and processing complex in western Colorado

GEMSTONES AND SPECIMEN-QUALITY MINERALS

The Pretty Stuff

For centuries, human inclination toward natural objects of beauty has led optimistic explorers to search for little pieces of Earth that are aesthetically appealing. Now, as in earlier days, people will pay dearly for rare and beautiful rocks and minerals. More than thirty different varieties of these gems and ornamental stones are known to occur in Colorado.

The most famous and historically valuable gems are diamonds. In 1975, diamonds were discovered in Colorado. The Kelsey Lake Diamond Mine, near the Wyoming border north of Fort Collins, began producing diamonds on a commercial scale in 1996—the only commercial producer in the United States. In 1997 the mine produced a yellow stone weighing 28.3 carats, the fifth largest diamond ever found in the United States. Weighing in at 16.87 carats after cutting and polishing, it became the largest faceted diamond ever produced in the United States.

The State Mineral is rhodochrosite, a manganese carbonate mineral found in eighteen of Colorado's counties. Red rhodochrosite crystals from the Sweet Home Mine near the town of Alma in Park County are prized all over the world for their exceptional size, color, and quality. The

Colorado's state mineral, Rhodochrosite (red), on tetrahedrite from Sweet Home Mine

finest specimens command prices up to $100,000. Ironically, early miners, intent on making a fortune from silver, probably threw away or crushed many fine specimens of rhodochrosite.

Aquamarine, the State Gem of Colorado, is a clear, blue variety of beryl. Mount Antero and nearby Mount White in Chaffee County are two of the best places in the world to collect specimens of this beautiful mineral. The collecting areas are near the 13,000-foot elevation.

Colorado's state gem, Aquamarine, from Mount Antero

Colorado diamonds "in the rough"

Wire gold from Breckenridge

Above: Crystalline gold from Breckenridge

Below: Smoky quartz and amazonite from Crystal Peak

Other notable gem-quality minerals that have been found in Colorado include amazonite, garnet, topaz, tourmaline, lapis lazuli, quartz crystal, smoky and rose quartz, amethyst, turquoise, peridot, sapphire, and zircon. Agate, chalcedony, and jasper are found in many places. Alabaster, a fine-grained, compact variety of gypsum used to make elegant vases and other decorative items, is quarried in the foothills northwest of Fort Collins and at a new quarry south of Carbondale.

STRIKING IT RICH AND STRIKING A NERVE

Do you own the mineral rights under your home? Most people in Colorado don't, and most people don't realize that they don't. There are many different "property rights" for a given piece of ground. One person may own the surface rights (usually the homeowner). Another may own the water rights. Another person or company may own the oil and gas rights. Another person may own the coal rights, another the rights to metals, and yet another the rights to geothermal resources.

Depending on how these various rights were legally set out when the original splits occurred, one "right" will usually have primacy over another. Mineral rights owners commonly have the legal right to develop the minerals on or under the property. Although they must compensate the surface owners for damages to their surface, the mineral owner does have the right to disturb the surface to retrieve the mineral resource, even though the surface owner may not want it disturbed. This can create ill feeling on both sides. Homeowners may feel betrayed because they didn't know this was the situation. The mineral owners know they have the legal right and may think the homeowners should have done their homework before making the biggest investment of their lives.

It is an understatement to say that recovery of mineral resources is not widely viewed as an aesthetically pleasing process. Mines and oil wells or drill rigs are usually not considered improvements to the landscape. They can be noisy, dusty, stinky, and create traffic. Whole communities have been known to march *en masse* to pound the desks of local officials demanding that they "do something" when a mining or petroleum operation threatens views and other cherished community amenities.

Beleaguered, extractive industry representatives point out that people can't expect to have houses, cars, sports equipment, and gasoline without allowing recovery of the resources that make it all possible. The public attitude, they contend, is uninformed at best and hypocritical at worst. To buttress their arguments, they cite development projects that have had to work through strident opposition of community members in order to retrieve valuable commodities—commodities, they point out that are essential to the very communities who resist their recovery.

One such example is in the suburban foothills west of Denver, where there are deposits of high-quality Precambrian metamorphic rock suitable for use as crushed stone aggregate used to build roads and make concrete. Aggregate producers trying to create quarries faced vocal opposition from foothill residents, many of who had large new houses and long commutes to work, lifestyles that consume great amounts of many resources.

Industry representatives also cite the citizens of Crested Butte, a town built on mining that now serves as a tourist destination for skiers and mountain bikers, for their reaction to the establishment of a molybdenum mine. A mining company wanted to work a world-class deposit at Mount Emmons a few miles west of town in the 1970s. The discovery of the deposit was significant because molybdenum is such a rare metal and there are very few mines in the world. Ironically, the metal is used to strengthen steel and is an ingredient in many mountain bike frames. The mining company representatives were perplexed when the town's many mountain bike-riding residents vehemently opposed the project.

Citizens and homeowners can describe examples of locations where mine operators did not exercise care and caution in the development or closure of a mine. The state has many locations where old-style mining, conducted before modern reclamation methods were required, left an eyesore or an environmental problem. There are no easy answers to the problem. The extractive resources industry has a right to recover resources, and citizens have a right to enjoy private and public property. Our society needs materials produced in mines and oil and gas fields. Additionally, we need to develop those operations while keeping the environment safe from degradation. The trick is to find the balance.

Colorado's Water

Why is water in a book on geology? Because geology guides the movement of water, both on the surface and in the subsurface, and provides storage capacity for water in the rocks. Water is also the primary geological agent in the evolution of landforms.

Rio Grande River in the San Luis Valley, with Sierra Blanca in background. This water has 1,250 more miles to go before reaching its destination in the Gulf of Mexico.

Colorado River near its headwaters in Rocky Mountain National Park, this water has 1,000 more miles to go before reaching its destination in the Gulf of California.

Here's a fact that may come in handy in your next game of geographic trivia: What state is known as the birthplace of rivers? Given that this is a book about Colorado, you can guess the answer, but the reasons why may surprise you. Few people know that in Colorado water comes almost entirely from precipitation *within* the state, or that all but one of Colorado's rivers flow out of the state providing water to eighteen other states. A number of the country's major rivers—including the North and South Platte, Rio Grande, Arkansas, and Colorado—begin as thousands of smaller tributary streams in the mountains of Colorado. Two raindrops falling only inches apart on Colorado's Continental Divide may end up in two different oceans, (i.e., the Atlantic or the Pacific).

Despite the number of rivers it births, Colorado as a whole is water deficient with an annual statewide average precipitation of only seventeen inches. The piedmont and plains areas lying east of the Rocky Mountains, where 90 percent of the population lives, receive a miserly twelve to sixteen inches annually, which is also all the plateau and valley regions of western Colorado can expect. Even worse, the San Luis Valley, a major agricultural area, receives only seven to twelve inches of rainfall each year.

Fortunately, this isn't the whole story. In contrast, and to the deep chagrin of highly populated areas where water is in desperately short supply, high mountain areas receive between twenty-five and forty inches of annual precipitation. Some areas receive as much as sixty inches and enjoy a luxurious surplus of this essential resource.

A semi-arid climate, a topography that interacts with meteorological systems causing sudden and sizable changes in the distribution of moisture, and rapid evaporation of whatever precipitation actually hits the ground makes predicting how much water will be available to state citizens from year to year dicey. More important than the amount of precipitation available, however, is where it falls. The major problem involving water in Colorado is how to move it from areas where it is abundant and under-utilized to areas where demand has outpaced supply. Complicated irrigation canals and water supply systems to deliver water where it was needed were developed soon after the state began to be settled. To keep up with demand in the eastern part of the state, large amounts of water have been diverted from the western slope of the Rockies to eastern flowing streams through ditches and tunnels. There are thirty-four separate trans-mountain diversion projects in Colorado.

Two large water diversion projects are the Colorado–Big Thompson Project that moves water from Grand Lake through a tunnel under the high mountains of Rocky Mountain National Park, and the Dillon Reservoir project that moves water twenty-three miles through a tunnel under the Continental Divide. Engineering geologists played a prominent role in each of these projects. They located sites for dams and tunnels in geologic formations that would not leak or fail so they could collect and transport a maximum amount of water during the high-runoff spring season for distribution through the rest of the year.

As the projects progressed, hundreds of new reservoirs were created to store irrigation water east of the mountains. Although water levels vary with the season, many of the reservoirs provide opportunities for water sports and recreation for the burgeoning population. An added bonus for geologists was that the tunnels and bores necessary to construct the systems gave them an opportunity to study the structure of the interior of the high mountains and helped improve their interpretation of Earth history in this interesting region.

Availability of water is so important to the state that water rights are spelled out in the Colorado Constitution. In this state, surface-water rights are governed by the Doctrine of Prior Appropriation. An easy way to remember the gist of the law is "first in time, first in right." Simply put, the person who first put water in a stream system to beneficial use has rights senior to others who began using the water later. Because of their value, water rights have become real property and can be bought and sold. Although the state needs water to fuel its growth and has an abundance of rivers, it is not free to use all of the water flowing through it. Downstream states have legal claims on much of the water originating in Colorado.

GROUNDWATER

In addition to water in streams and lakes (surface water), water under the surface (groundwater) is a valuable resource. If you dig a hole, you generally find that the soil is relatively dry near the surface. Eventually, however, your hole will reach a depth where the ground is saturated with water. The upper surface of the saturated zone is the water table. Below the water table, groundwater completely fills the void spaces between soil and rock particles, and the fissures and fractures in rock.

Groundwater provides 16.5 percent of Colorado's total water use of about 14 billion gallons per day. To many remote users who are far from public water supply systems, groundwater is more than important, it is essential. As is true of surface water, most groundwater in Colorado is used to irrigate agricultural crops.

Groundwater is usually of very good quality for most uses. Mechanical processes in the unsaturated zone filter bacteria and sediment from water as it percolates into the subsurface. Chemical processes can also remove impurities and add desirable (and sometimes undesirable) minerals to the water.

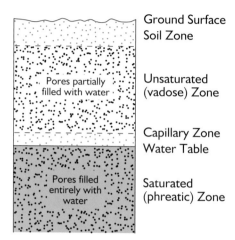

This diagram illustrates that as you dig deeper and deeper, eventually you will reach a place where all of the pore spaces are saturated with groundwater, the water table. Groundwater migrates through the tiny holes in the rock or sediment.

Aquifers are permeable rocks in the saturated zone that transmit water freely. They act both as reservoirs and as conduits for groundwater storage and flow. The USGS identifies seven principal aquifers or aquifer systems in Colorado: South Platte Aquifer, Arkansas Aquifer, High Plains Aquifer, San Luis Valley Aquifer System, Denver Basin Aquifer System, Piceance Creek Basin Aquifer, and the Leadville Limestone Aquifer of west-central Colorado.

In addition to the seven principal aquifers, there are many less extensive and less productive aquifers that are locally very important. Sufficient water for a single household may be relatively easily obtained from wells in the bedrock aquifers, even the crystalline Precambrian rocks of the mountains. However, as households multiply, creating increased demands from many wells, a local aquifer with very limited storage that was sufficient for a few users, may be quickly depleted or polluted.

Although the principle use of groundwater is for irrigation, groundwater is also used to meet nearly all livestock and rural domestic water needs. By 1980, the High Plains Aquifer was tapped by approximately 15,600 wells. Since the only source of recharge to the High Plains Aquifer is precipitation, which averages from fourteen to eighteen inches per year, these wells are drawing water from storage.

An important but often ignored reality is that groundwater is a finite resource and becomes mined or depleted if use exceeds the aquifer recharge rate. Some aquifers are recharged consistently by streams and occasional floods. In contrast, deeper bedrock aquifers must rely on the small fraction of precipitation that falls on the recharge area and moves into the water table below. These older and deeper aquifers are more extensive and have attained their stored water slowly over thousands of years. If people withdraw water from an aquifer faster than nature replenishes the aquifer, the water table drops. If this continues, wells go dry or become too costly to operate.

Several residential areas located in the crystalline (granitic) rocks of the Colorado Mountains have wells that dry up. This results from drought conditions or too many closely spaced wells withdrawing water too fast. The lesson is simple, groundwater withdrawals that greatly exceed the sustainable yield (recharge rate) of the aquifer cannot be maintained for long.

HOT SPRINGS

Another fabulous water resource in Colorado is its hot springs. Throughout Colorado's history, the benefits of relaxing in the warm, soothing waters of this state have been appreciated. Native Americans used natural hot springs for health and spiritual purposes, and early European settlers enthusiastically immersed themselves, just as many people do today.

Colorado has ninety-three known geothermal areas including natural springs, augmented natural springs, and wells. Colorado's geothermal resources are classified as low temperature—less than the boiling point of water—and are used for recreational bathing, aqua-culture, minor space heating, and heating green houses.

Probably the most famous hot springs are those around Glenwood Springs. The springs issue from Mississippian limestone and are the saltiest in the state. Scientists believe that the high salinity of the springs around Glenwood Springs is related to dissolution of salt in the nearby Pennsylvanian evaporites.

A cold winter morning accents Glenwood Hot Springs Resort's warm waters in the pool, which was constructed in 1888 and is much the same as it was more than 100 years ago. The Big Spring, which services the pool, has a variable flow rate, but was measured at more than twenty-two gallons per second at a temperature of 124 degrees Fahrenheit in 1978.

Other places in Colorado, including Idaho Springs, Pagosa Springs, Steamboat Springs, Hot Sulfur Springs, Ouray, Mount Princeton, and Poncha Hot Springs, are well-known for their popular spas.

Recently, hot springs have been tapped for uses more exotic than soaking tired muscles. Two sites in the state are used to raise fish (aqua-culture). The hottest springs in the state are around Mount Princeton. With temperatures hitting 185 degrees Fahrenheit, the springs provide heat for a resort, residences, and a greenhouse before flowing into Chalk Creek. About two miles downstream, Chalk Creek has cooled to a temperature that the Colorado Division of Wildlife considers ideal for use in its trout-rearing unit. Generally, it takes a fingerling somewhere between eighteen and twenty-two months to become a ten-inch trout. The fingerlings allowed to grow up basking in warm water, however, accomplish the same feat in twelve months flat.

The Hooper Well near the small town of Hooper in the San Luis Valley is another unique aqua-culture site. The Hooper Well, it turns out, is just the right temperature to raise African Perch, and alligators. It is curious to realize that having warm water enables these diverse and unlikely tropical endeavors to thrive in one of the coldest regions of the state.

Wetlands are abundant in Rocky Mountain National Park's Kawunee-chee Valley

WETLANDS

Wetlands are areas that contain seasonally or perennially saturated soils and specialized, water-loving plants. Once reviled as insect-infested swamps, wetlands are now valued as being ideal for wildlife habitat, groundwater storage, flood attenuation, streambank stabilization, heavy metal and sediment retention, and recreation. Less than 2 percent of Colorado's land is made up of wetlands; the state has lost about half its wetlands in the 150 years since its pioneer days.

In Colorado, wetlands take many forms including snowmelt depressions, hillside seeps, slope wetlands, peatlands, wet meadows, marshes, groundwater flats in closed basins, and riparian wetlands along floodplains. Colorado's largest wetland area is found in the San Luis Valley at the San Luis Lakes near the Great Sand Dunes. This closed-basin, groundwater flat contains many playas and marshes and is an important wildlife management area.

In recent years, people have begun to understand the important link between geology and wetlands. Colorado's wetlands exist because of topography created by past and present geologic events and processes. They depend on off-site sources of water and water pathways that are controlled by topography and subsurface geology, and the hydrodynamics of the watershed. In sum, the health and viability of the state's wetlands depends not only on managing the wetlands themselves, but also on managing and maintaining the water sources and pathways within the surrounding watershed areas.

GEOLOGY AND WATER QUALITY

As snowmelt and rain enter Colorado streams and percolate into the ground, the water picks up particles and dissolves components of rocks. Ordinarily, this natural process does not affect the water quality enough to be of concern. In some areas, however, elements in the rocks are dissolved in high enough concentrations to adversely affect living organisms.

Metals in Colorado's mountains sometimes seep into streams from abandoned mines and waste rock piles. In other streams, high concentrations of metals occur naturally as a result of the geology of the area. The North Fork of the South Platte River in Park County, South Fork of Lake Creek in Lake County, and the upper Alamosa River

Areas of hydrothermal alteration, such as Red Mountain south of Independence Pass, naturally contribute metals and acidity to local watersheds. Aluminum and iron precipitates drop out of these springs as they flow at the surface.

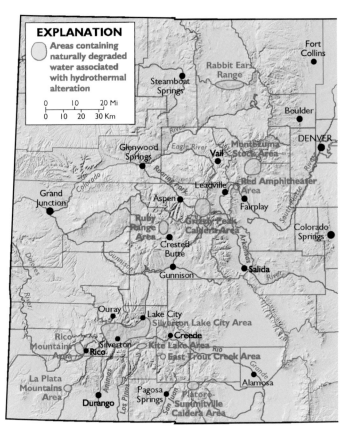

Map showing areas in Colorado where CGS scientists have identified natural geologic conditions contributing metals and acidity to streams. Although many of these areas are in areas of mining activity, the natural degradation of the streams is often greater than the effects of mining.

in Conejos County are prime examples of natural degradation. Many times, metals in Colorado's headwater streams are derived from both natural and mine-induced sources.

Acid rock drainage occurs when water and oxygen interact with metal-sulfide minerals, such as pyrite, producing sulfuric acid that dissolves metals and carries them into groundwater and streams. Natural acid rock drainage has been active in Colorado for at least thousands, possibly millions, of years.

High salinity concentrations can also be a concern for water quality. In 1973, high salinity in the Colorado River Basin from Colorado to California was considered serious enough to warrant an amendment to an international treaty between Mexico and the United States. The amendment set standards for the total amount of dissolved solids the water could carry as it flowed into Mexico. The federal government initiated some large projects to reduce the salinity of the river and its tributaries.

The causes of high salinity in the Colorado River water are varied, but geology is an important factor. Thick deposits of ancient sea salt lie under parts of the river basin in Colorado. Groundwater and surface water continually dissolve this salty layer, contributing to salinity downstream. Recent geologic mapping by the Colorado Geological Survey indicates that this dissolution process has been going on for several million years and, over this span of time, as much as 540 cubic miles of salt deposits may have been carried away by streams. Natural hot springs, such as those near Glenwood Springs and Dotsero, carry especially large amounts of dissolved material and aggravate the salinity problems of the Colorado River.

Selenium is an essential nutrient required for human health, but in high concentrations it can be toxic. The toxicological effects and environmental occurrences of the element selenium have been studied since the 1930s when it was discovered that high concentrations in pasture grasses caused disease and death in cattle and horses. Since that time, selenium has also been found to disrupt the embryonic development of birds and fish. Selenium occurs in higher than normal concentrations in several Colorado streams; it predominantly comes from water flowing back into the river after irrigating agricultural areas with certain types of soil.

The Upper Cretaceous marine shales in Colorado generally contain higher concentrations of selenium than other rocks in Earth's crust. Dissolved selenium is picked up by streams as they flow across the shales. As the water flows, some of it evaporates making the selenium in the streams even more concentrated. Areas of concern in Colorado are the Gunnison River Basin/Grand Valley area, Pine River Basin, and Middle Arkansas River Basin. Certain reaches of these streams have selenium concentrations above the national maximum contaminant level of five micrograms per liter. Many national, state, and local organizations are trying to address this problem.

Naturally occurring radioactive elements can occur in groundwater in Colorado and are a health risk in high concentrations. Radon, radium, and uranium are found in small amounts in most rocks. If rocks that have greater than normal concentrations of these elements form an aquifer, the local groundwater may contain unacceptably high levels of radioactivity. An expensive case in point is in the town of Castle Rock, which drilled a $100,000 water-supply well and encountered excessive amounts of radioactivity in the water.

Geology plays an important role in water quality. Some water quality issues in Colorado have their foundation in the natural interaction between water and rock. This interaction can produce poor water quality independent of other influences, but existing problems are often exacerbated when people become part of the equation. The geology and water quality of an area should be assessed before any development takes place—this will alert planners to potential problems. Careful planning can help preserve water quality.

Colorado's Geologic Hazards

Colorado is often misjudged. Early explorers, battered and beaten by the rugged terrain, reported back to Washington, D.C. that there was no point in considering settlement in Colorado—the place was absolutely unfit for human habitation.

In 1806, desperate, weary, and woefully unequipped for a winter storm, Zebulon Pike gave up his only attempt to climb the peak that now bears his name, predicting that no one would ever reach the summit. He was wrong. By 1891, it was possible for corseted ladies in full paint and powder to ascend Pike's Peak on a cog railway. These days, you can assault the mountain without ever leaving the comfy confines of the family sedan and be back in Colorado Springs in time for lunch.

Modern homesteaders roar into the state by the SUV-load and stake claim to sites on the flanks of sleeping volcanoes and build homes in valleys that once formed the oozy bottom of ancient seas, aware only that the view is spectacular. In today's Colorado, it is easy to assume that the forces that created the attractive landscapes are now quaint and quiescent relics of the distant past. Wrong, again!

Those who approach the ruggedly beautiful vistas of the state should do so with caution and be armed with knowledge of potential hazards lurking underfoot. Some of Colorado's hazards are relatively obvious. Rocks, soil, and snow perched at the top of steep slopes, given any chance, will try to join their counterparts settled in the valleys below. When much of the state was still wilderness

My basement broke! What happened? Earthquake? Landslide? Hydro-compactive soil? Uh-oh, swelling soil. Now I wish I hadn't watered so close to the foundation.

and inaccessible, the tendency of this or that mountain to drop some of its bulk down its sides, if observed at all, was of little concern. The few hardy souls who made this land their home were usually wise enough to understand this phenomena and stay out of the way.

As the state's population grew, homesteaders fearlessly hacked away at the land as they built homes and roads. To enhance their new property, homeowners trapped the water in reservoirs and streams and proceeded to plant yards, gardens, and farms. Others, drawn by dreams of fabulous wealth, enthusiastically burrowed underground chasing elusive veins of gold and silver.

As a result of all this frenetic activity, people began to discover some of Colorado's less endearing surprises, which include swelling and collapsing soils, landslides, debris flows, rockfalls, avalanches, floods, and earthquakes. As more and more people came to Colorado, formerly inaccessible or undesirable land became ripe for development. The need for assessment and mitigation of geologic hazards risks prompted the Colorado Legislature to pass laws requiring developers in unincorporated areas of the state to conduct a detailed analysis of geologic hazards and to

submit the analysis for review by the Colorado Geological Survey. Many municipalities, while not required by law to do so, also voluntarily request land-use reviews by the state geological survey. If property owners, including the state entities, are informed of any problems associated with a particular area and undertake appropriate mitigation, development of industry, residences, and recreational facilities can take place without exposing people to unacceptable risks. Plus, it's usually much cheaper to mitigate the hazard before it has caused damage.

SWELLING SOILS/HEAVING BEDROCK

Innocuous as it sounds, Colorado's most significant geologic hazard is swelling soil—soil laced with layers of various clays. These clays cause more property damage than any other natural hazard. Bentonite and montmorillonite (weathered volcanic ash) clays underlie many populated areas of Colorado. They can expand up to 20 percent in volume when exposed to water and exert up to 30,000 pounds of force per square foot, more than enough to break up any structure they encounter. One zip code in a Denver suburb has the dubious distinction of suffering more annual monetary loss from swelling soil than any other in the nation.

Heave ridges in street and sidewalk in a southwest suburb of Denver where swelling clay layers are turned up on end near the mountain front. The clay layers expand when they get wet, but the intervening layers don't.

Where the claystone layers turn up on end near the foothills, the effects of swelling are intensified and the phenomenon is called heaving bedrock, which causes heave ridges. These ridges cause roads to ripple and fully grown highway engineers to weep with frustration. A number of roads in Jefferson County west of Denver, including C-470, near Bowles Avenue and Wadsworth Boulevard, have heave ridges, a telling sign that extraordinary precaution is needed to prevent structural catastrophes in these areas.

Sound building techniques can prevent swelling-soil damage to homes, but it is crucial that builders follow these techniques faithfully. Even after homes are built, homeowners must be careful to keep water away from the foundation to avoid it being damaged. CGS geologists wrote a booklet about this hazard for homeowners that has sold more than 175,000 copies and won several national awards.

Problems with clay layers, of course, are not restricted to family residences. Public buildings and schools throughout the state have been affected as well. Buildings at the University of Southern Colorado campus and the Colorado State Prison have suffered millions of dollars worth of damage from swelling soil.

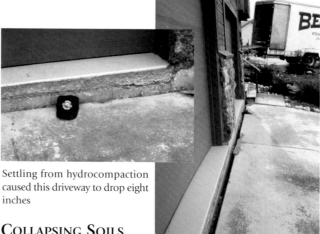

Settling from hydrocompaction caused this driveway to drop eight inches

COLLAPSING SOILS

Collapsing, or hydrocompactive, soils are swelling soils' evil twin. These soils are fairly strong when dry, but can lose 15 percent of their volume when wet and cause the ground to sink several feet. Hydrocompactive soils form when materials such as wind-blown loess, slope wash, debris flows, and mudflows are not well compacted when deposited and later become compacted by the addition of water. Hydrocompaction also occurs in soils formed from evaporites, which is prevalent in central and southwestern Colorado.

SUBSIDENCE

The surface of the land can also sink, or even open up, when material is removed from underground by mining or natural processes. Subsidence occurs over old landfills, poorly engineered construction fills, or improperly filled clay pits. Students of the Colorado School of Mines in Golden who lived in housing constructed over badly compacted clay mines gained a special insight into the problem when their closet doors stuck shut and cupboard doors perversely swung open no matter how firmly they were closed.

Empty, near-surface coal mining tunnels, the result of extensive mining in Colorado Springs, eastern Jefferson County, and from Boulder eastward to I-25 may collapse in response to increased surface weight or construction activity. Called subsidence pits, these depressions dot the urban corridor along the Front Range and may be eight to ten feet deep.

Swelling soil damage at the University of Southern Colorado

CGS geologist standing in collapse depression caused by hydrocompaction and possibly deeper dissolution of evaporites

Surface collapse into an abandoned, underground coal mine

LANDSLIDES

Many of Colorado's mountainous areas are susceptible to landslides, as are the sides of mesas and river valley slopes on Colorado's plains. Rocks, soils, and artificial fills on slopes are constantly subject to the downward pull of gravity. When slopes yield, landslides occur. Landslides have a hummocky appearance. Scarps commonly form at the top of a landslide where the earth is pulling away and mounds form at the bottom where material is piling up.

Colorado Springs in particular has problems with subsidence as a result of abandoned mine tunnels that snake under much of the city. For city officials, a memorable illustration of the problem was watching a swimming pool built over one of these old mines begin to sink. Concrete was quickly pumped into the old shafts to provide support, but the pool at that point was about a foot and a half lower than originally intended.

Subsidence can also occur when rocks, generally limestones and evaporites, dissolve under the surface. If the dissolving rocks are far enough from the surface to prevent collapse, the resulting caves are often an interesting addition to the area. If the dissolution occurs closer to the surface, sinkholes can form. Subsidence hazards and sinkholes plague Colorado south and east of Glenwood Springs and along the eastern side of the Front Range. Horsetooth Reservoir near Fort Collins suffered so much damage from evaporite dissolution that it had to be emptied in 2000 for repair.

Destruction of a brand-new home in Grand Junction by a landslide caused by the Colorado River eating away at the hillside and creating gravitational instability

Landslides can move so slowly as to be imperceptible or quickly enough to embarrass a track star. Landslides are triggered naturally by saturation of the ground (either by heavy rains and/or ground seepage), change in slopes caused by erosion or construction, and earthquakes. Saturation and resultant landslides can also be triggered by dedicated gardeners incessantly pouring water on lawns to keep them golf-course green, and by ambitious property owners or road builders bent on excavating unstable slopes.

Although some landslides may be dormant for years, they may be reactivated by changes such as development of the land. Each year, buildings and roads are constructed in areas prone to landslides. In some cases, these risks can be reduced by mitigation techniques, such as grading, landscaping, and slope stabilization. In many cases, the wisest

course for a property owner to take is admitting defeat and finding a more docile bit of land, ideally one that will stay where it was originally found.

Periodic landslides along the flanks of North and South Table Mountains near Golden have played havoc with railroad tracks, irrigation ditches, and roads. In one area, successive layers of pavement have been laid one on top of another to keep a street "up to grade." Asphalt there is estimated to be thirteen feet thick. At nearby Green Mountain, a landslide destroyed three large homes in 1998, following several years of above-average precipitation. Colorado Springs and Boulder also contain landslide-prone areas. Wet winters and heavy rains triggered damaging landslides in Colorado Springs in 1995 and again in 1998. Current estimates for damage and mitigation in the Colorado

Photograph and topographic map showing the famous Slumgullion landslide near Lake City, which is continuously moving in its upper part. The slide is a little over four miles long and began moving at least 2,000 years ago. The active part of the slide is moving as much as twenty-one feet per year.

MESSAGES IN STONE

Springs neighborhoods are at $75 million. On the Western Slope, landslides have caused damage in numerous communities including Grand Junction and Telluride.

Large, earthflow-type landslides pose a threat to the state's highways in certain locations. Examples include the Slumgullion landslide near Lake City that dammed the Lake Fork of the Gunnison River to form Lake San Cristobal. This huge landslide continues to inch down the mountain. As the slide moves, the pine trees in its path are tilted to peculiar angles before they are finally ripped from their roots and float along with the slide. Bumpy, frequently patched roads close to McClure Pass are the signature of the Muddy Creek landslide. Other landslide areas include the Dowd's Junction landslide near Minturn, and a large, unnamed landslide complex along the southern approach to Douglas Pass.

The DeBeque Canyon landslide poses a major threat to I-70, the Colorado River, and an arterial rail line. This slide is still moving, and is closely monitored by Colorado Geological Survey and the Colorado Department of Transportation. Although it is unlikely, it could fail catastrophically in the future.

One of the few redeeming features of earth movements is that many of Colorado's ski areas feature slopes that are mantled with landslide deposits, forming legendary ski slopes, bowls, and hummocky glades.

In 1991 in the San Juan Mountains, a massive volume of rock failed catastrophically above West Lost Creek. Ten and a half million cubic yards of debris was triggered by heavy summer rains and melting of residual ice on the north-facing slope. The debris ran up the opposite slope to a height of nearly 200 feet. The slide is estimated by an eyewitness to have traveled a half mile in about thirty seconds.

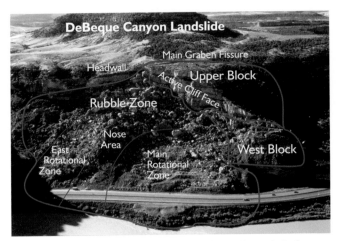

This large slide in DeBeque Canyon is moving at about six inches per year and threatens I-70, which cuts across its toe.

House built in a flood plain at the toe of a landslide

Rockfall in DeBeque Canyon during the olden days

CGS geologist identifying large rock that needs to be stabilized in Glenwood Canyon above I-70

ROCKFALL

Where rocks are exposed along steep slopes or cliffs, rockfall is inevitable. A few small rocks tumbling down may not be cause for concern, but a house-sized boulder careening toward your new home in the valley is something else entirely.

The state's transportation department is acutely aware of the hazard posed by falling rocks and has allocated a substantial portion of its budget to engineering mitigation techniques with the aim of protecting both highways and adjacent residential communities. Highway travelers should watch for fallen rocks as they drive through canyons and beneath steep slopes on many of the state's highways, especially after storms or during periods of freezing and thawing. Many mountain highways of Colorado have hazard signs where rockfall commonly occurs.

Rocks are a particular problem in beautiful Glenwood Canyon. About $11 million has been spent trying to protect this highway, which has been identified as one of the most rockfall-prone sections of the entire interstate highway system.

Large rockfall along I-70 west of Denver in 1973, fortunately no travelers were involved

Left: Vail resident contemplating what might have happened if she had been home when a large boulder crashed into her bedroom

Above: Massive barrier constructed to prevent recurrence of boulders crashing into the condominium below

Below: This rock fell from above prior to construction of this home. A similar boulder striking the home would do considerable damage to residents and building alike. However, the Big Thompson flood devastated this canyon before another large boulder could fall.

Avalanches

Avalanches, defined by state statute as a geologic hazard, are a product of weather and geology. As they slide, avalanches scour debris from gullies and slope faces and cut chutes in the rock. Each year, an estimated 20,000 avalanches slide in the mountains of Colorado, making "the Winter Sports Capital of the Country" second only to Alaska in the sheer number of avalanches and first, regrettably, in the number of avalanche-related deaths. This is because of a combination of factors including type of snow and climate, access to avalanche areas, and the number of people in avalanche areas. Approximately 100 avalanches each year involve people and property, and cause injury, death, or damage. In addition, approximately 300 avalanches block Colorado highways each winter.

Many avalanches are small, perhaps running as little as 100 to 200 feet, and are relatively harmless, though they could still bury an unsuspecting skier. Large storms produce large avalanches. In the spring, an entire winter's accumulation of snow can tear loose with catastrophic results. These huge avalanches may release a mass of snow that is five to ten feet deep—more than 500,000 cubic yards of snow—over an area of twenty football fields. Sometimes falling more than 3,000 vertical feet, these avalanches attain speeds greater than 100 miles per hour, and destroy everything in their paths. Avalanche frequency can vary greatly, with some paths releasing perhaps twenty times a year and others, only once in twenty years.

Skiers and snowmobiles can, and often do, trigger avalanches. Not all survive. In the decade of the 1990s, an average of six people died annually in Colorado in avalanches, and direct property damage was approximately $100,000. The Colorado Avalanche Information Center, a section of the Colorado Geological Survey, works hard to minimize the impact of avalanches on the lives of Coloradoans and visitors through a program of forecasting and education, and they have succeeded. During the decade of the 1990s, the number of deaths per 100,000 population actually decreased in Colorado. In contrast, the death rate from avalanches increased markedly in Alaska, Utah, and Montana.

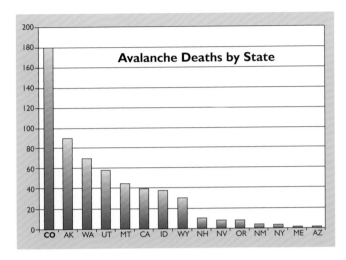

Avalanche Deaths by State

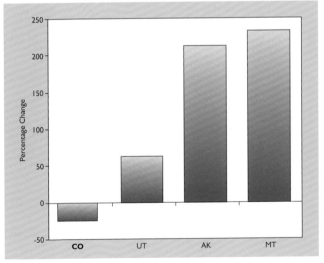

Avalanches flattened this truck (left) and nearly flattened the Airstream trailer (above)

Avalanche awareness courses have had an impact over the last decade in reducing deaths in Colorado.

Damage from 1965 Denver flood on the South Platte River

FLOODS

Floods, in Colorado's semi-arid climate? You bet. As they say, "When it rains, it pours." Fort Collins residents can testify to the inundation that occurred from more than fourteen inches of rain being dumped on their city in thirty-one hours in 1997. That is the amount typically received in a whole year. Floods perennially threaten much of the state because of the high topographic relief of the drainage basins, torrential spring thaws, and intense summer thunderstorms. Early pioneers often chose to build highways, railroads, and towns in canyons close to riverbanks. Predictably, flooding was among the first natural hazards of which new citizens became painfully aware. Large floods devastated Denver, Pueblo, and other Colorado cities.

Colorado's most expensive flood was probably the "Flood of 1965" in the South Platte River Basin south of Denver. This flood caused $508 million worth of damage and drowned six people. The losses can be attributed to the failure to realize the significance of the South Platte drainage routes and flood plains. Homes, shopping centers, and many other buildings occupied—and still occupy—land that has been intermittently flooded for many years.

The following description of the 1965 flood published by the USGS may help to convey some warning to residents or potential residents of the South Platte valley and other river valleys in Colorado. East Plum Creek, referenced in the article, is normally an innocuous creek flowing northward along I-25 south of Castle Rock. The article reads as follows:

"The morning of June 16 was most pleasant, but conditions changed rapidly shortly before noon. At about 2:00 P.M., a dense mass of clouds descended and concealed the top of Dawson Butte, seven miles southwest of Castle Rock; and the little light remaining faded until it was dark black and frightening, according to some people.

The deluge began, not only near Dawson Butte, but also Raspberry Mountain, six miles to the south, near Larkspur. The rain came down harder than any rain the local residents had ever seen, and the temperature dropped rapidly until it was cold. The quiet was shattered by the terrible roar of wind, rain, and rushing water. Then the thudding of huge boulders, the snapping and tearing of trees, and the grinding of cobbles and gravel increased the tumult. The small natural channels on the steep slopes could not carry the runoff, so water took shortcuts, following the line of least resistance. Creeks overflowed, roads became rivers, and fields became lakes—all in a matter of minutes.

The flow from glutted ravines and from fields and hillsides soon reached East and West Plum Creeks. The combined flow in these creeks has been described as awesome, fantastic, and unbelievable; yet none of these superlatives seem adequate to describe what actually occurred. Large waves, high velocity crosscurrents, and eddies swept away trees, houses, bridges, automobiles, heavy construction equipment, and livestock. All sorts of debris and large volumes of sand and gravel were torn from the banks and beds of the streams and were dumped, caught, plastered, or buried along the channel and flood plains downstream. A local resident stated, 'The banks of the creek disappeared as if the land was made of sugar.'

The flood reached the South Platte River and the urban areas of Littleton, Englewood, and Denver at about 8:00 P.M. Here the rampaging waters picked up house trailers, large propane storage tanks, lumber, and other flotsam and smashed them against bridges and structures near the river. Many of the partly plugged bridges could not withstand the added pressure and washed out. Other bridges held, but they forced water over approach fills, causing extensive erosion. The flood plains carried and stored much of the floodwater, which inundated many homes, businesses, industries, railroad yards, highways, and streets.

The flood peak passed through Denver during the night, and the immediate crisis was over by morning; but those in the inundated areas were faced with a Herculean task. The light of day revealed the nature of the destruction—mud in every nook and cranny, soggy merchandise, warped bowling alleys, drowned animals, the loss of irreplaceable possessions, to name a few types. The colossal cleanup job, which would take months, began."

Another devastating flood occurred in 1976 when 139 lives were lost in Big Thompson Canyon between Estes Park and Loveland. This flood resulted from a single thunderstorm that dropped about twelve inches of rain in a few hours. A flash flood roared through the narrow canyon, rising to as much as fourteen feet above normal river levels. There was little or no chance of escape for many of the canyon's residents and visiting tourists.

Geological studies by the Colorado and United States Geological Surveys show that other canyons along the Front Range have suffered similar devastating floods and can experience them again. Do not ignore signs in Colorado's canyons that warn "Climb to Safety." Casualties result when people try to out-drive a flood surge. Better to get wet than lose your life.

The Big Thompson Flood of 1976 wiped out homes and lives with sudden rising waters.

Debris Flows and Alluvial Fans

Debris flows are a common hazard in many hillside areas of Colorado. Heavy rainfalls commonly trigger flash floods on steep slopes. These torrents pick up anything in their paths and may contain more solid material than liquid. They tear into the hillsides and deposit accumulated material when they reach flatter ground, creating an alluvial fan. A superlative alluvial-fan deposit can be viewed in Horseshoe Park in Rocky Mountain National Park. This deposit formed on a single day in July 1982, when the Lawn Lake Dam burst, sending a huge surge of water and debris down the Roaring River Valley. Although alluvial fans may appear to be attractive building sites, it is prudent to remember that another debris flow may follow the same course that created this area. If it does, your new home may well end up a bit farther down the valley and flatter than you had planned.

Recurring debris flows on alluvial fans worry residents of mountain towns such as Ouray, Glenwood Springs, Marble, and Georgetown, and for good reason. A debris flow destroyed the newly constructed Primate House at the Cheyenne Mountain Zoo in Colorado Springs in 1965. Farther up Cheyenne Mountain, debris flows forced the temporary closure of the North American Air Defense (NORAD) facility. Three people were killed by the flooding downstream from the zoo.

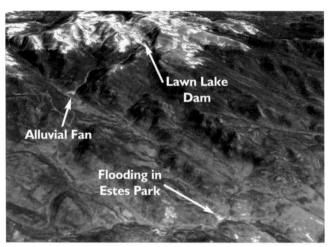

The Lawn Lake Dam burst high in a glacial valley and scoured the stream channel down to Horseshoe Valley where it killed some campers and deposited a large fan. The rushing waters surged downstream to Estes Park causing flooding and property damage.

Downstream flooding in Estes Park from the dam burst in Rocky Mountain National Park, July 15, 1982

Forest fires, to which the arid west is vulnerable, are terrible in themselves. Not as well understood is that danger and damage often continue after the flames are out. When a forest fire occurs in the mountains, most people do not automatically think, "call a geologist," but it's not really such a bad idea. Forest fires remove stabilizing vegetation and when combined with the right (or wrong) geology, commonly set up conditions that lead to major flooding and debris flows during subsequent heavy rains. Flooding following a fire in the Front Range southeast of Denver resulted in tremendous problems with silt deposited in the Strontia Springs Reservoir on the South Platte River. Millions of dollars were spent repairing the damage to this

This alluvial fan formed in Rocky Mountain National Park when a dam burst in 1982, killing three people.

Debris flows are very damaging when houses and humans are in their path such as this one in Glenwood Springs. Their paths, however, are quite predictable.

Debris flow that buried Highway 109 near Carbondale

Westbound lanes of Interstate 70

Debris flows are common in the Georgetown area and have closed lanes of I-70 on more than one occasion

reservoir, and scientists estimate that 463,000 cubic yards of material were deposited in the reservoir in the two-and-a-half years after the fire. This represents only a fraction of the total soil eroded from the fire area. After a large fire, CGS geologists quickly notify local jurisdictions of those areas that are most susceptible to debris flows.

Although newspapers across the country reported the tragic deaths of fourteen firefighters battling the Storm King Mountain fire west of Glenwood Springs in 1994, little national attention was given to the dangerous debris flows in the fire's aftermath. Two months after the fire, torrential rains triggered debris flows that reached I-70 in fifteen places. Thirty cars traveling on the highway were engulfed, trapped, or overturned by the mud. A fourteen-year-old eyewitness who was trying to help a stranded motorist reported:

"As I approached the vehicle, the mud flow was about ankle deep, but the flow increased rapidly and soon it was thigh deep. The chunks of rocks and trees that I could feel, knocked me off my feet and I was carried by the mudflow across the road, over the edge, and into the Colorado River. As I moved over the rock ledge in the churning mud, rocks, logs, and debris hit under my chin. I was actually relieved to reach the river where I could move and the water washed away the muck from the flow."

The lucky young man escaped with his life, but his jaw was broken in three places.

Humans and Geology

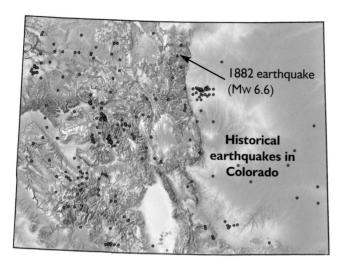

1882 earthquake (Mw 6.6)

Historical earthquakes in Colorado

EARTHQUAKES

An earthquake is simply the vibrations caused by fault movement. The bigger the movement, the bigger the earthquake. Because the mountains are still rising in Colorado, earthquakes will continue to accompany the faulting that enables them to grow. More than 500 earthquakes have been recorded throughout Colorado in historic times. The largest historical earthquake was in 1882 located near Estes Park and is estimated to have been magnitude 6.6. This earthquake knocked out power in Denver and was felt as far away as Salt Lake City.

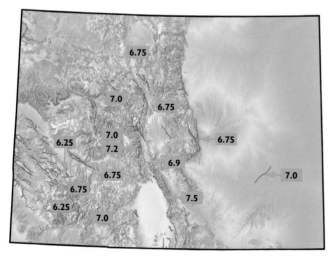

"Maximum credible earthquake" is a term used by engineering geologists for structural design purposes. It refers to the largest earthquake that can reasonably be expected to occur on a particular fault. This map shows the magnitude of the maximum credible earthquakes assigned to various faults around the state.

Geologists from the Colorado Geological Survey compiled a list of ninety faults in the state that have moved in the recent geologic past. The geologic characteristics of several of these faults indicate they are capable of generating large earthquakes in the future. Earthquake studies in Colorado are still in their infancy compared to many other states, and much remains to be done regarding the risk of a major earthquake affecting populated areas of the state.

Human-Induced Earthquakes

Colorado is a world-renowned laboratory for human-induced earthquakes. The saga began in Denver in the early 1960s at the Army's Rocky Mountain Arsenal. In response to environmental concerns about leaving liquid waste in surface ponds, the Army drilled a well and pumped the waste two miles underground, where it supposedly would not bother anyone. Not too long afterward, Denver began experiencing earthquakes that, with time, grew larger and larger.

A Denver geologist soon correlated the volumes of liquid being pumped down the well with the frequency of earthquakes, and pronounced that the pumping was causing the earthquakes. The Army said he was crazy. Most geologists in Denver said he was crazy. However, studies by government scientists confirmed that he was indeed correct.

Three of the earthquakes reached magnitudes greater than 5.0 in 1967. A geologist now on staff with the Colorado Geological Survey remembers being terrified as a child when one of the earthquakes threw her out of her bed and pushed her home slightly off its foundation. This earthquake did a million dollars worth of damage in northeast Denver. Fortunately, the Army abandoned the idea of pumping waste fluid underground, and the problem was solved. Fifteen years later, the allegedly crazy Denver geologist won a cash award from the National Council on Environmental Quality for his work in showing that the pumping and earthquakes were related.

The USGS thought the notion of creating earthquakes was intriguing, and located an oil field in northwestern Colorado where fluid injection seemed to be causing a

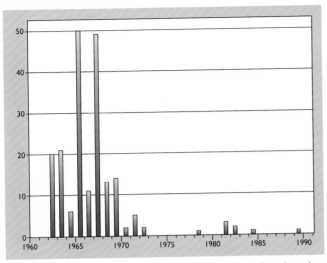

Induced earthquakes near Denver: Annual number of earthquakes recorded in the vicinity of the Rocky Mountain Arsenal Well. Most scientists consider those in the 1960s to have been induced by fluid injection two miles underground.

CONCLUSION

Despite the potential for landslides, falling rocks, and earthquakes, Colorado is a beautiful state, and one to which people will continue to be drawn. As we accumulate knowledge and develop techniques to mitigate dangers, chances are we can all live successfully with the rocks that surround us. It may be wise, however, to keep historian Will Durant's caveat in mind, "Civilization exists by geological consent, subject to change at a moment's notice."

series of small earthquakes. They convinced the oil company to let them perform an experiment to see whether they could turn the earthquakes off and on. When the geoscientists stopped the fluid injection, the micro-earthquakes dropped from around fifty per day to around one per day. When they got the injection pressure back up to where it had been, the number of earthquakes jumped to previous levels. When they stopped injecting again, the earthquakes stopped. This was the first experiment in the world where earthquakes were intentionally turned off and on by geoscientists.

In 2000, the Bureau of Reclamation began injecting fluid into the ground in western Colorado as part of an effort to reduce the salinity of water entering the Colorado River. Because of these two previous experiments, they expected that they would generate earthquakes, but only small ones. Indeed they did generate more than 3,500 small earthquakes, but, when one earthquake exceeded magnitude 4.0 in 2000, they decided it was time to cut back.

Geologists beware! The Colorado Department of Transportation, apparently unsatisfied with Nature's handiwork, has begun creating fake outcrops such as this one on Highway 24 west of Woodland Park.

Further Reading and Resources

(*Titles available from the Colorado Geological Survey)

General—Popular

Chronic, Halka and Felice Williams. *Roadside Geology of Colorado*, second edition. Missoula: Mountain Press Publishing, 2002.*

Foutz, Dell R. *Geology of Colorado Illustrated*. Dell R. Foutz, 1994.

Hopkins, Ralph Lee and Lindy Birkel Hopkins. *Hiking Colorado's Geology*. Seattle: The Mountaineers Books, 2000.*

Johnson, Kirk R. and Robert G. Raynolds. *Ancient Denvers: Scenes from the Past 300 Million Years of the Colorado Front Range*. Denver: Denver Museum of Nature & Science, 2001.

Johnson, Kirk R. and Richard K. Stucky. *Prehistoric Journey: A History of Life on Earth*. Boulder: Roberts Rinehart Publishers, 1995.

Murphy, Jack A. *Geology Tour of Denver's Buildings and Monuments*. Denver: Historic Denver, 1995.*

———. *Geology Tour of Denver's Capitol Hill Stone Buildings*. Denver: Historic Denver, 1997.*

Osterwald, Doris B. *Rocky Mountain Splendor: A Mile by Mile Guide for Rocky Mountain National Park*. Lakewood: Western Guideways, 1989.*

Raup, Omer B. *Colorado Geologic Highway Map*. Denver: Colorado Geological Survey, 1991.*

Raup, Omer B. and Curt Buchholtz. *Geology Along Trail Ridge Road, Rocky Mountain National Park, Colorado: A Self-Guided Tour for Motorists*. Guilford: Falcon Publishing, 1996.*

Taylor, Andrew M. *Guide to the Geology of Colorado*. Cataract Lode Mining Company, 1999.

General—Technical

Geologic Atlas of the Rocky Mountain Region, United States of America. Rocky Mountain Association of Geologists, 1972.

Kent, Harry C. and Karen W. Porter, editors. *Colorado Geology*. Rocky Mountain Association of Geologists Symposium, 1980.

Tweto, Ogden. *Geologic Map of Colorado*. USGS, 1979.*

———. *Geologic Cross Sections Across Colorado* (five geologic sections that supplement the Geologic Map of Colorado). USGS, 1983.*

Fossils

Jenkins, Janice L. and John T. Jenkins. *Colorado's Dinosaurs*. Denver: Colorado Geological Survey, 1993.*

Voynick, Stephen M. *Colorado Rockhounding: A Guide to Minerals, Gemstones, and Fossils*. Missoula: Mountain Press Publishing, 1995.*

Economic

Ayers, W. B. and W. R. Kaiser, editors. *Coalbed Methane in the Upper Cretaceous Fruitland Formation, San Juan Basin, New Mexico and Colorado*. New Mexico Bureau of Mines and Mineral Resources, 1994.*

Davis, M. W. and R. K. Streufert. *Gold Occurrences of Colorado*. Denver: Colorado Geological Survey, 1990.*

Eckel, Edwin B. et al. *Minerals of Colorado*. Denver: Denver Museum of Nature & Science and Fulcrum Publishing, 1997.*

Mitchell, James and Robin Nordhues, editor. *Gem Trails of Colorado*. Baldwin Park: Gem Guides Book, 1997.

Parker, B. H. *Gold Panning and Placering in Colorado: How and Where.* Denver: Colorado Geological Survey, 1992.*

Pearl, Richard M. *Colorado Gem Trails and Mineral Guide.* Athens: Ohio University Press, 1972.*

Wray, Laura, Christopher J. Carroll, John W. Keller, and James A. Cappa. *Colorado Mineral and Mineral Fuel Activity, 2000.* Denver: Colorado Geological Survey, 2001.*

Hazards

Amuedo and Ivey. *Coal Mine Subsidence and Land Use in the Boulder–Weld Coalfield, Boulder and Weld Counties, Colorado.* Denver: Colorado Geological Survey, 1975.*

Kirkham, Robert M. and William P. Rogers. *Colorado Earthquake Information 1867–1996.* CD-ROM. Denver: Colorado Geological Survey, 2000.*

Noe D. C., C. L. Jochim, and W. P. Rogers. *Guide to Swelling Soils for Colorado Homebuyers and Homeowners.* Denver: Colorado Geological Survey, 1997.*

Rogers, W. P. and Robert M. Kirkham. *Contributions to Colorado Seismicity and Tectonics: A 1986 Update.* Denver: Colorado Geological Survey, 1986.*

Caves

Parris, Lloyd E. *Caves of Colorado.* Boulder: Pruett Publishing, 1982.

Rinehart, Richard J. *Colorado Caves: Hidden Worlds Beneath the Peaks.* Englewood: Westcliffe Publishers, 2001.*

Water

George, Deborah Frazier. *Colorado's Hot Springs,* second edition. Boulder: Pruett Publishing, 2000.*

Topper, Ralf. *Ground Water Atlas of Colorado.* Denver: Colorado Geological Survey, 2003.*

Neubert, John T. *Naturally Degraded Surface Waters Associated with Hydrothermally Altered Terrain in Colorado.* CD-ROM. Denver: Colorado Geological Survey, 2000.*

Pearl, Richard H., Ted G. Zacharakis, Frank N. Repplier, and Kevin P. McCarthy. *Bibliography of Geothermal Reports in Colorado.* Denver: Colorado Geological Survey, 1981.*

Field Guides

Donnell, John R., editor. *Geological Road Logs of Colorado.* Rocky Mountain Association of Geologists, 1960.

Epis, Rudy C. and Robert J. Weimer, editors. *Studies in Colorado Field Geology.* Colorado School of Mines, 1976 Annual Meeting Geological Society of America, November 8–11, 1976.

Lageson, David R., Alan P. Lester, and Bruce D. Trudgill, editors. *GSA Field Guide 1, Colorado and Adjacent Areas.* 1999.

Mountain Geologist. Guidebook Issue, Part I, Vol. 9, Nos 2–3. Rocky Mountain Association of Geologists, 1972.

Mountain Geologist. Guidebook Issue, Part II, Vol. 9, Nos 2–3. Rocky Mountain Association of Geologists, 1972.

Weimer, Robert J. and John D. Haun, editors. *Studies in Colorado Field Geology.* Geological Society of America, Rocky Mountain Association of Geologists, Colorado Scientific Society, 1960.

Image Credits

Photographs and Images by Source

INDIVIDUALS

Bruce Bryant: 87 UL LR

Peter Birkeland: 78 LR

Braddock: 42 U

Dave Bunk: 124 UR LR, 125

James A. Cappa: 5 UL, 29 UL, 59 LL, 60 UL

Chris Carroll: 115, 116 UL

Cater: 43 UR

Dave Catts: All digital elevation models

William A. Cobban: 23 LR

Howard Coopersmith: 124 LL

Emmett Evanoff: 22 LR, 58 UL

Mike Frasier: 97 UR

David A. Gonzales: 5 L, 31 LL ML UR MR, 80 LL

Francisco Gutierrez: Cover, ix L, 38 U, 87 UR, 97 LR, 145 UL

Wallace R. Hansen: 34 UL, 35 UL UR, 40 UL, 63 ML, 72 UR, 94 LL, 102 R, 138 LR, 143 LR

David Harris: 95 LR, 96 LR

P. J. Hasselbach: 3 LM, 4 inset, 10 UR, 32 R, 39 M L, 40 MR, 64 C, 69 LR, 70, 79 UL, 103 ML, 113 UL UR, 117

Ralph Lee Hopkins: vii U L, 8 LR, 9 LL, 18 LL, 24 LR, 30 UL ML, 68 UR, 77 L, 93 L, 99 L

Kirk R. Johnson: 23 U, 27 LL, 71 UL

John A. Karachewski: Inside cover, title page, x L, 3 LL, 6 UR, 8 LL, 9 R, 17 LR, 32 L, 34 L, 58 R, 60 LR, 61 R, 63 U MR LR LL, 64, 67 UR, 74 UR, 81 L, 82 ML, 90 R, 91 L, 93 UL, 104 R, 111 LR, 112 UR, 135 LR

Allen Karsh: 12 LL

John W. Keller: 30 LR, 93 MR, 109 U

G. Kent Keller: 29 LR

Katie KellerLynn: 22 LL, 54 UR, 79 ML

Kirkham: 146 UL

Brendan LaMarre: 96 UL

Trina LaMarre: 96 UL

Tim Lane: 140

Sandra J. Lindquist: vi, 19 UR, 20 UL ML, 90 LL, 91 UR LR

Peter Lipman: 12 UR

Richard Madole: 89 UR

Vincent Matthews: iii, v, viii U L, ix U, x UL, xi, xii, xiv, 3 U, 4 U LR, 5 UR M, 6 UL M L, 7 LL C, 8 UR, 9 UL, 10 LL, 11 LL LR, 12 LR, 15, 16, 17 U LL, 19 UL, 21 LR, 26 LR, 27 LR, 28 UR, 31 LR, 33, 35 LL, 36 UL LR, 38 L, 39 U, 40 L, 41, 42, 43 MR, 44 U ML LR, 45, 48, 49, 51, 52, 54 LL, 55, 56 LL, 57 L, 62, 63 LL MR LR, 65 UR, 66, 67 LL, 69 UL, 71 UR, 72 LL LR, 73, 74 UL L, 75, 76 UR, 79 LR LL, 80 UL MR, 81 LR, 82 UR, 83, 84, 85 LR, 88, 89 UR, 90 UL, 92 L, 94 UL UR LL, 95 UL, 98, 99 U, 100 LL UR UL ML, 101 U LL, 103 LL LR, 104 UL, 105, 106 LL, 107 U LR ML, 109 LL LR, 110, 112 UL LL, 113 LL, 114, 119 LR, 120, 121, 122, 126, 129, 135 LR, 139 L, 144 UR LL, 146, 147

Meierding: 78 LR

Bill Middlebrook: 63 LM

Karen Morgan: 12 UL

Matt Morgan: 13 L, 14 U M

Eric Nelson: 37 LR

John Neubert: 131 LL UR

David C. Noe: 7 UR, 63 U, 65 L, 86 R, 106 UL, 133, 134 LL, 137 LR

Brian Penn: 4 LR

Lloyd Parris: 95 LL, 96 LL

Richard M. Pratt: 79 UR

Jerry Roberts: 141 UR UL

W. P. Rogers: 106 LR, 145 UR, 146 UL

Eric Route: 30 UR

Matthew Sares: 131 LL

Larry Scott: x UR, 11 U, 13 U, 43 UR, 48 UL, 101 LR

Jeff Scovil: 124 UR LR, 125

Kenneth Lee Shropshire: 37 LL

James Soule: 108 L

Jack Stanesco: 20 UR LL

Ray Troll: 27 U

Tweto: 3 UR, 7 C, 48 L, 51 UL, 55 UR, 62 UR, 65 UR, 71 UR, 75 UL, 77 UL, 89 UR, 109 LL LR

Joe Tucciarone: 22 UR

Van Horn: 73 M

Varnes: 88 ML

Theodore Walker: 18 LL, 19 L, 20 UC LR

Mark Wark: 53 UR, 99 M, 102 L

Jonathan White: 7 UL, 68 LL, 86 L, 89 LR, 97 LL, 107 LL, 123 UR, 134 UR, 137 LL, 138 UR, 139 UL UR, 145 MR

Beth Widmann: 43 LL, 110 UL

Jason Wilson: xiii, 53 LL

Laura Wray: 113 UL, 116 LR

INSTITUTIONS

American Soda: 123

Colorado Department of Transportation: 138 UL

Colorado Geological Survey: 135 UL UR, 106 LR

Colorado Historical Society: 29 UR, 97 UL, 118

Denver Museum of Nature & Science: 21 UR, 24 UL, 25 UR LL, 26 UL, 71 LR

Florissant Fossil Beds National Monument: 21 ML LL

National Aeronautics and Space Administration: xi, 14 L

Nevada Bureau of Mines and Geology: 36 LL MR, 37 ML MR, 38 MR LR

Rocky Mountain National Park: 1, 3 LR, 30 LL MR, 78 LL UR, 82 LR, 85 UR LL, 130, 144 MR

Stagner, Inc.: 25 LR

United States Geological Survey: xi, 7 LR, 28 LR, 46, 47, 76, 87 LL, 100 LR, 119 U, 127, 142, all geologic maps and derivatives are from the USGS

United States Army: 29 M

United States Forest Service: 137 UR

University of Colorado Department of Geological Sciences: 2 U L, 28 LL, 44 LL, 92 U, 103 UR, 104 LL, 108 U, 136

Photographs and Images by Page

Cover: Francisco Gutierrez

inside cover, title page: John A. Karachewski

iii: Vincent Matthews

v: Vincent Matthews

vi: Sandra J. Lindquist

vii: LL Ralph Lee Hopkins, R Ralph Lee Hopkins

viii: L Vincent Matthews, UR Vincent Matthews

ix: L Vincent Matthews, R Francisco Gutierrez

x: UL Vincent Matthews, UR Larry Scott, LR John A. Karachewski

xi: UR Vincent Matthews

xii: L Vincent Matthews

xiii: M Jason Wilson

xiv: L Vincent Matthews

1: LR Rocky Mountain National Park

2: UL, LL University of Colorado Department of Geological Sciences

3: UR Vincent Matthews, geology from Tweto, 1979, LL John A. Karachewski, M P. J. Hasselbach, LR Rocky Mountain National Park

4: U Vincent Matthews, LL Brian Penn, LR Vincent Matthews

5: UL James A. Cappa, UR Vincent Matthews, M Vincent Matthews, L David A. Gonzales

6: UL Vincent Matthews, UR John A. Karachewski, ML Vincent Matthews, L Vincent Matthews

7: UL Jonathan White, UR David C. Noe, M Vincent Matthews, geology from Tweto, 1979, LL Vincent Matthews, LR William Henry Jackson

8: UR Vincent Matthews, L John A. Karachewski, ML Ralph Lee Hopkins

9: UL Vincent Matthews, ML Ralph Lee Hopkins, R John A. Karachewski

10: L Vincent Matthews, R P. J. Hasselbach

11: U Larry Scott, LL Vincent Matthews, LR Vincent Matthews

12: UL geology from Steven, 1975 and Lipman 2000, LL Allen Karsh, UR Peter Lipman, LR Vincent Matthews

13: UR Larry Scott, L Matt Morgan

14: UL Matt Morgan, MR Matt Morgan, L NASA

15: U Vincent Matthews

16: U Vincent Matthews, ML Vincent Matthews, MR Vincent Matthews, L Vincent Matthews

17: UR Vincent Matthews, LL Vincent Matthews, LR John A. Karachewski

18: LL Ralph Lee Hopkins, LR Theodore Walker

19: UL Vincent Matthews, UR Sandra J. Lindquist, MR Theodore Walker

20: UL Sandra J. Lindquist, UC Theodore Walker, UR Jack Stanesco, ML Sandra J. Lindquist, LL Jack Stanesco, LR Theodore Walker

21: UR Denver Museum of Nature & Science, L Florissant Fossil Beds National Monument, LR Vincent Matthews

22: UR Joe Tucciarone, LL Katie KellerLynn, LR Emmett Evanoff

23: UC Kirk R. Johnson, LR William A. Cobban

24: UL Denver Museum of Nature & Science, LR Ralph Lee Hopkins

25: UR Denver Museum of Nature & Science, MR Stagner, Inc., LL Denver Museum of Nature & Science

26: UL Denver Museum of Nature & Science, LR Vincent Matthews

27: U Ray Troll, LL Kirk R. Johnson, LR Vincent Matthews

28: UR Vincent Matthews, LL University of Colorado Department of Geological Sciences, LR USGS

29: UL James A. Cappa, UR Colorado Historical Society, MR United States Army, LR Kent G. Keller

30: UL Ralph Lee Hopkins, ML Ralph Lee Hopkins, LL Rocky Mountain National Park, MR Eric Route, MR Rocky Mountain National Park, LR John W. Keller

31: ML David A. Gonzales, LL David A. Gonzales, UR David A. Gonzales, MR David A. Gonzales, LR Vincent Matthews

32: L John A. Karachewski, R P. J. Hasselbach

33: UL Vincent Matthews, R Vincent Matthews

34: UL W. R. Hansen, L John A. Karachewski

35: UL W. R. Hansen, UR W. R. Hansen, LL Vincent Matthews

36: UL Vincent Matthews, MR Nevada Bureau of Mines and Geology, LL Nevada Bureau of Mines and Geology, LR Vincent Matthews

37: ML Nevada Bureau of Mines and Geology, MR Nevada Bureau of Mines and Geology, LL Kenneth Lee Shropshire, LR Eric Nelson

38: U Francisco Gutierrez, MR P. J. Hasselbach, L Vincent Matthews, LR P. J. Hasselbach

39: U Vincent Matthews, LR P. J. Hasselbach

40: UL Wallace R. Hansen, L Vincent Matthews, MR P. J. Hasselbach

41: U Vincent Matthews, LL Vincent Matthews, LR Vincent Matthews

42: U Vincent Matthews, geology from Braddock et al., 1970, L Vincent Matthews

43: UR Larry Scott, MR Dave Catts, LL Beth Widmann

44: U Vincent Matthews, ML Vincent Matthews, LL University of Colorado Department of Geological Sciences, LR Vincent Matthews

45: U Vincent Matthews, MR Vincent Matthews, LR Vincent Matthews

46: ML USGS, MC USGS, LR USGS

47: C geology from Scott et al., 2001 and USGS

48: UL Vincent Matthews, geology from Scott et al., 2001, UR Vincent Matthews, L Vincent Matthews, geology from Tweto, 1979 and USGS

49: LR Vincent Matthews

51: UL Vincent Matthews, geology from Tweto, 1979, R Vincent Matthews

52: UR Vincent Matthews

53: LL Jason Wilson, UR Mark Wark

54: LL Vincent Matthews, UR Katie KellerLynn

55: LL Vincent Matthews, UR Vincent Matthews, geology from Tweto, 1979

56: LL Vincent Matthews

57: L Vincent Matthews

58: UL Emmett Evanoff, UR John A. Karachewski

59: LL James A. Cappa

60: LR John A. Karachewski, UL James A. Cappa

61: R John A. Karachewski

62: UR Vincent Matthews, geology from Tweto, 1979, L Vincent Matthews

63: U David C. Noe, ML W. R. Hansen. LL Vincent Matthews, MR Vincent Matthews, LR Vincent Matthews, LC Bill Middlebrook

64: UR John A. Karachewski

65: UR Vincent Matthews, L David C. Noe

66: UR Vincent Matthews, geology from Tweto, 1979, MR Vincent Matthews, L Vincent Matthews

67: UR John A. Karachewski, LL Vincent Matthews

68: UR Ralph Lee Hopkins, LL Jonathan White

69: UL Vincent Matthews, LR P. J. Hasselbach

Image Credits

70: C P. J. Hasselbach, LR David C. Noe

71: UL Kirk R. Johnson, UR Vincent Matthews, geology from Tweto, 1979, LR Denver Museum of Nature & Science

72: UR Wallace R. Hansen. LL Vincent Matthews, LR Vincent Matthews

73: UR Vincent Matthews, M Vincent Matthews, geology from Van Horn, 1956

74: UL Vincent Matthews, UR John A. Karachewski, L Vincent Matthews

75: UL Vincent Matthews, geology from Tweto, 1979, UR Vincent Matthews

76: R USGS

77: UR Vincent Matthews, geology from Tweto, 1979, L Ralph Lee Hopkins

78: UR Rocky Mountain National Park, LL Rocky Mountain National Park, LR glaciers from Meierding and Birkeland, 1980

79: UR Richard M. Pratt, LR Vincent Matthews, ML Katie KellerLynn, LL Vincent Matthews

80: UL Vincent Matthews, MR Vincent Matthews, LL David A. Gonzales

81: L John A. Karachewski, UR Rocky Mountain National Park, LR Vincent Matthews

82: ML John A. Karachewski, UR Vincent Matthews, LR Rocky Mountain National Park

83: UL Vincent Matthews and Dave Catts, L Vincent Matthews

84: UL Vincent Matthews, ML Vincent Matthews, LL Vincent Matthews, UR Vincent Matthews, MR Vincent Matthews, LR Vincent Matthews

85: LL Rocky Mountain National Park, UR Rocky Mountain National Park, LR Vincent Matthews

86: LL Jonathan White, R David C. Noe

87: UL Bruce Bryant, ML USGS, UR Francisco Gutierrez, LR Bruce Bryant

88: ML from Varnes et al., 1989, LL Jack Stanesco, LR Vincent Matthews

89: UR geology from Madole, 2003 and Tweto, 1979, LR Jonathan White

90: UL Vincent Matthews, LL Sandra J. Lindquist, R John A. Karachewski, MR NASA

91: L John A. Karachewski, UR Sandra J. Lindquist, LR Sandra J. Lindquist

92: U University of Colorado Department of Geological Sciences, L Vincent Matthews

93: L Ralph Lee Hopkins, UR John W. Keller

94: UL Vincent Matthews, LL Wallace R. Hansen, UR Vincent Matthews, LR Vincent Matthews

95: UL Vincent Matthews, LL modified from Parris, 1973, LR David Harris

96: UL Brendan and Tricia LaMarre, LL modified from Parris, 1973, LR David Harris

97: UL Colorado Historical Society, LL Jonathan White, UR Mike Frasier, LR Francisco Gutierrez

98: U Vincent Matthews, MR Vincent Matthews

99: UL Vincent Matthews, MR Mark Wark, LL Ralph Lee Hopkins

100: UL Vincent Matthews, LL Vincent Matthews, UR USGS, MR Vincent Matthews, LR William Henry Jackson and USGS

101: U Vincent Matthews, LL Vincent Matthews, LR Larry Scott

102: L Mark Wark, MR Wallace R. Hansen

103: ML Vincent Matthews, LL Vincent Matthews, UR University of Colorado Department of Geological Sciences, LR Vincent Matthews

104: UL Vincent Matthews, LL University of Colorado Department of Geological Sciences, R John A. Karachewski

105: UL P. J. Hasselbach, ML Vincent Matthews, LL Vincent Matthews, UR Vincent Matthews, MR Vincent Matthews, LR Vincent Matthews

106: UL David C. Noe, LL Vincent Matthews, LR W. P. Rogers

107: U Vincent Matthews, ML Vincent Matthews, LL Jonathan White, LR Vincent Matthews

108: UR University of Colorado Department of Geological Sciences, L Jim Soule

109: UL John W. Keller, LL Vincent Matthews, geology from Tweto, 1979, LR Vincent Matthews, geology from Tweto, 1979

110: UL Vincent Matthews, L Vincent Matthews, UR USDA

111: LR John A. Karachewski

112: UL Vincent Matthews, LL Vincent Matthews, UR John A. Karachewski

113: UL from Wray, 2002, LL Vincent Matthews, UR CGS

114: M Vincent Matthews

115: LL Chris Carroll, UR Chris Carroll, MR Chris Carroll, LR CGS

116: UL Chris Carroll, LR Laura Wray

117: UR Matthew Sares

118: LR Colorado Historical Society

119: U USGS, LL CGS, LR Vincent Matthews

120: UR Vincent Matthews

121: ML Vincent Matthews, UR Vincent Matthews

122: LL Vincent Matthews, UR Vincent Matthews

123: UR Jonathan White, LR American Soda

124: LL Howard Coopersmith, UR Dave Bunk specimen, Jeff Scovil photo; LR Dave Bunk specimen, Jeff Scovil photo;

125: UL Dave Bunk specimen, Jeff Scovil photo, UR Dave Bunk specimen, Jeff Scovil photo, M Dave Bunk specimen, Jeff Scovil photo

126: LR Vincent Matthews

127: UL William Henry Jackson and USGS

129: UR Vincent Matthews

130: U Rocky Mountain National Park

131: LL geology from Neubert, 2000 and Larry Scott and P. J. Hasselbach, UR John Neubert

133: UL David C. Noe, LR David C. Noe

134: LL David C. Noe, UR Jonathan White

135: UL CGS, UR CGS, LR Vincent Matthews

136: L University of Colorado Department of Geological Sciences

137: UR United States Forest Service, LR David C. Noe, LL Jonathan White

138: UL Colorado Department of Transportation, UR Jonathan White, LR Wallace R. Hansen

139: UL Jonathan White, UR Jonathan White, L Vincent Matthews

140: L Tim Lane

141: UL P. J. Hasselbach, LL Jerry Roberts, UR Jerry Roberts, LR P. J. Hasselbach

142: U Warren B. Hamilton and USGS

143: UC Vincent Matthews, LR W. R. Hansen

144: LL Vincent Matthews, UR Vincent Matthews, MR Rocky Mountain National Park

145: UL Francisco Gutierrez, UR W. P. Rogers, MR Jonathan White

146: UL Vincent Matthews, geology from Kirkham and Rogers, 2000, LL Vincent Matthews

147: UL CGS, LR Vincent Matthews

Index